THE
BOOK OF
GARDEN FLOWERS

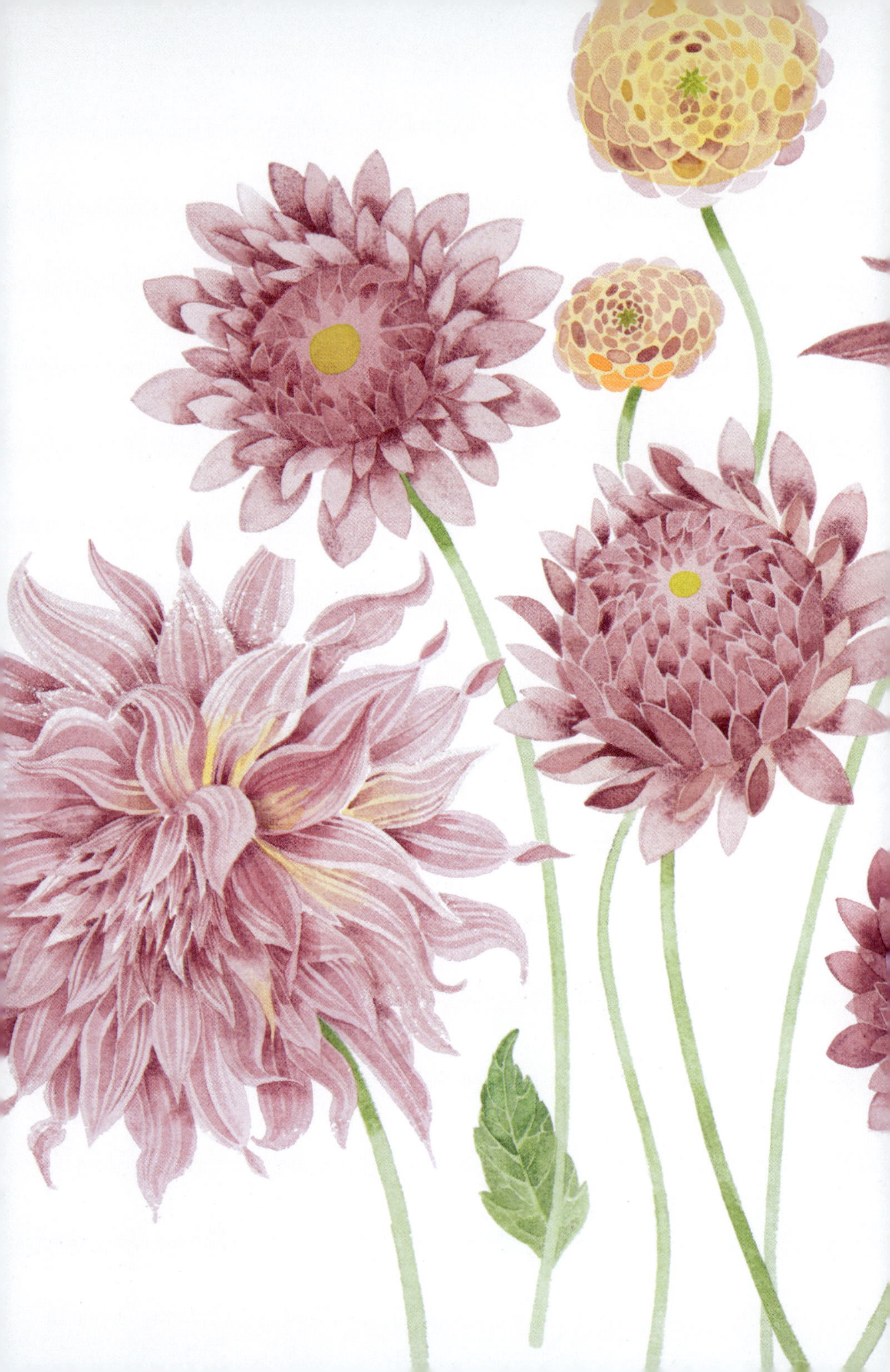

THE
BOOK OF
GARDEN FLOWERS

———

Written by Christopher Stocks

Illustrated by Angie Lewin

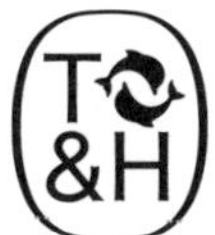

Frontispiece: *November, Last of the Dahlias* (detail)
Watercolour, 2022

First published in the United Kingdom in 2025 by
Thames & Hudson Ltd, 6–24 Britannia Street, London WC1X 9JD

First published in the United States of America in 2025 by
Thames & Hudson Inc., 500 Fifth Avenue, New York, New York 10110

The Book of Garden Flowers © 2025 Thames & Hudson Ltd, London

Text © 2025 Christopher Stocks

Foreword and artwork © 2025 Angie Lewin

Designed by Simon Lewin

EU Authorized Representative: Interart S.A.R.L.
19 rue Charles Auray, 93500 Pantin, Paris, France
productsafety@thameshudson.co.uk
www.interart.fr

A CIP catalogue record for this book is available from the British Library

Library of Congress Control Number 2024943772

ISBN 978-0-500-02707-3
02

Printed and bound in Slovenia by DZS-Grafik d.o.o.

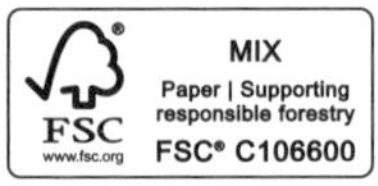

Be the first to know about our new releases,
exclusive content and author events by visiting
thamesandhudson.com
thamesandhudsonusa.com
thamesandhudson.com.au

CONTENTS

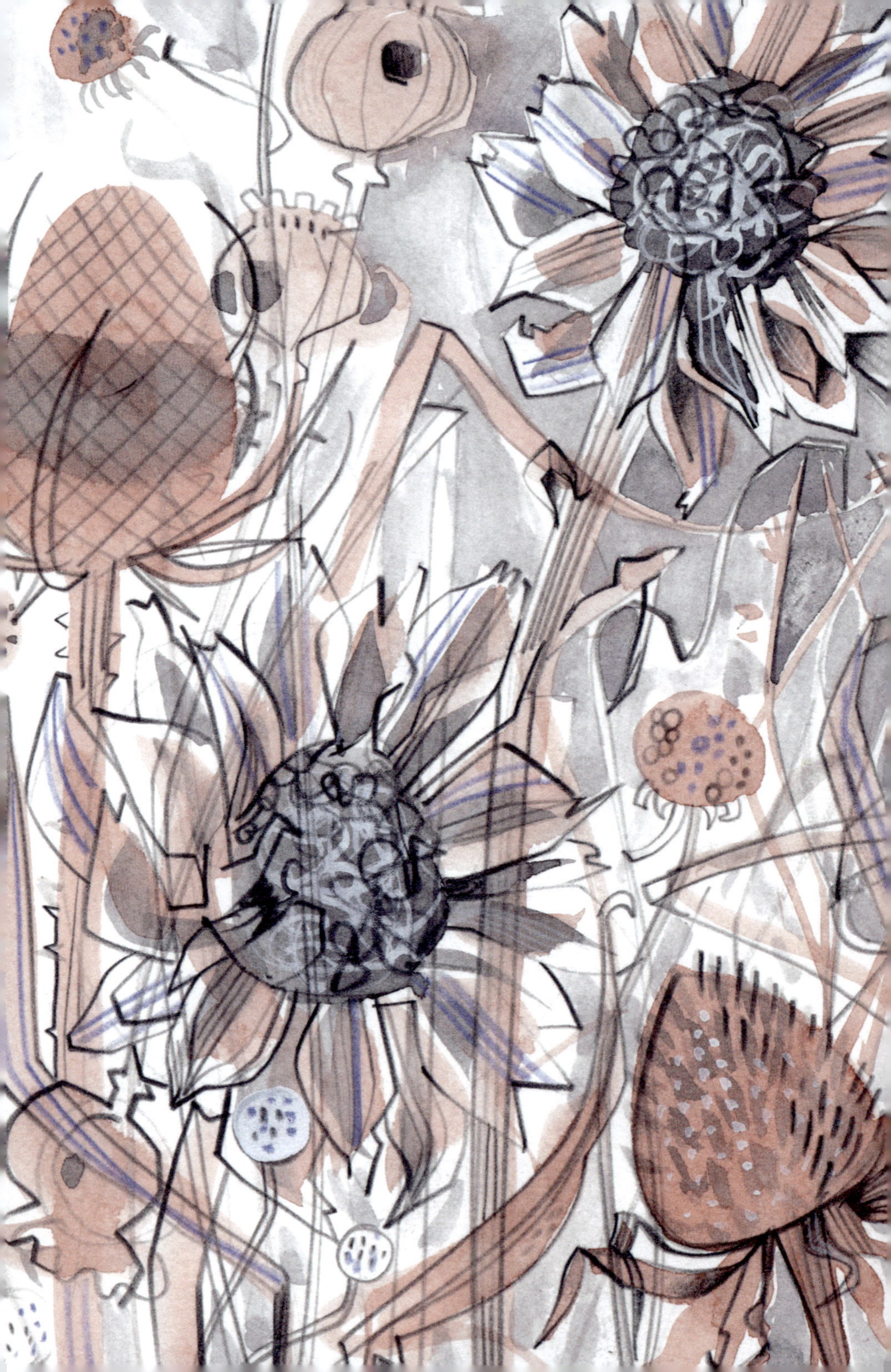

FOREWORD

A GROWING INSPIRATION

ANGIE LEWIN

My very earliest memories are of a garden. I believe I could sketch it even now: the sunken lawn surrounded by a low stone wall covered with succulents and candytuft, the rhubarb patch and rows of raspberries and gooseberries. There was a greenhouse full of tomatoes with their musty smell, and beds of rosebushes and dahlias edged with colourful bedding plants that my father would grow from seed or buy from the colour supplement of his Sunday newspaper. I spent much of my time there as a small child, making daisy chains, twirling buttercups and weaving grass stalks. When I was older, I would make drawings of the rows of broccoli and cabbages in the vegetable plot, and use the lawn to construct sculptures out of branches and found objects from the old sheds and greenhouse.

Sketch for *Autumn Garden, Norfolk* screenprint
Mixed media, 2013

I went on to study fine art at the Central School of Arts and Crafts in London, before finally moving into a flat with a small yard where I could have a few pots and grow herbs and tomatoes. It seemed the natural thing to do, something that had been missing from my life. Then, another move took me to a typical long north London garden. There was a gate through to our neighbour Maggie's garden, which was formal but softened by wild flowers. This became a significant influence on how I began to design and plant my own plot. The ends of the two gardens, at Maggie's suggestion, were linked by a band of white-stemmed birch trees that screened the ugly flats beyond. The way the trunks gleamed in front of the old, ivy-covered brick wall and emerged from an underplanting of acanthus and evening primrose was magical.

I soon wanted to learn more, so, in the mid-1990s, I studied for an RHS Certificate in Horticulture and then garden design at Capel Manor College in north London. The plant portrait sketchbooks I made at this time were the early inspiration for drawing and painting wild and garden flowers. Each week, we were given ten specimens to study, sketch their form and structure, and research their habitat, propagation and cultivation. It was at this point, as I looked so intently, that I realized how even a seemingly insignificant species can be as fascinating as the most exotic. I began to develop my botanical studies into linocuts, wood engravings, lithographs, screenprints and, later, textile and wallpaper designs.

I left London for North Norfolk in the late 1990s and moved to an old red-brick and flint cottage with a large garden. When we first visited, it had been uninhabited for a long time. Poppies and ox-eye daisies grew all the way up a narrow track to the cottage door. When I began to garden there, I loved the combination of wildness and cultivation, so I introduced plants that emphasized this quality. Experimenting with form, structure and colour in the garden is remarkably similar to composing a print or a painting. As gardeners, we introduce plants from disparate places into our gardens: a rose from China may sit alongside a *Verbena bonariensis* from South America, and soon, native yarrow and plantain will happily tuck themselves in around them. In my artwork, I often bring together the wild and the cultivated to celebrate the relationship between them.

Until a few years ago, I mainly sketched and developed prints from my time spent in the wilder landscapes of northern and west-coast Scotland and the North Norfolk coastline. My garden might appear only occasionally, represented by agapanthus, artichoke and echinops (globe thistle). Perhaps because of the coronavirus lockdowns in 2020 and 2021, my focus shifted more towards my Edinburgh studio and garden. I couldn't return to my home on Speyside, and my regular trips to North Norfolk and yearly stay on North Uist in the Outer Hebrides were not possible. Still-lifes, assembled in my studio from jars and pots of seedheads, lichened twigs and feathers, were now even more important than ever in evoking memories of these places. My collection of auriculas came into flower during the first lockdown of spring 2020, and I focused on these jewel-like plants while I was working on a solo watercolour exhibition at the Scottish Gallery later that year. I realized that the intense observation of plants, of which I was so aware when out walking and sketching in the countryside, was equally relevant in my small city garden.

Now that I live full-time in rural north-eastern Scotland, I grow dahlias, tulips and anemones in a Shetland 'Polycrub' to evoke memories of this urban garden. Other cultivated species thrive in sheltered parts of the garden close to the cottage and steadings. I enjoy the way these non-native plants associate with the wild flowers and grasses and the textured grey lichen-covered stone of the buildings and walls. Auricula flowers are almost childlike in their simplicity, but I can lose myself in their depth of velvety colour and the contrasting rosettes of their glaucous leaves. I try to capture these colours with washes of intense watercolour. The steely blue flowerheads of echinops punctuate my paintings just as they are dotted through my garden borders. In my still-life paintings I like to contrast intensely colourful, blowsy dahlias with chalky grey-green lichens and purple-brown twisted branches of heather, heightening the qualities of each.

This eclectic collection of plant portraits reflects my daily life as an artist and gardener. There are also plants, such as ranunculus and *Anemone coronaria*, which I choose to grow simply to paint and for cut flowers for the house. As is the case for many gardeners, some plants were gifts from friends or have travelled with me as I have moved house and garden over the years. They also reflect the progression of my work, evolving each time I redraw them or make a new watercolour or print, and emerging repeatedly in my work, just as they might self-seed in the garden.

Angie Lewin, Speyside, Scotland

Charleston Garden sketch
Watercolour, pencil and crayon, 2019

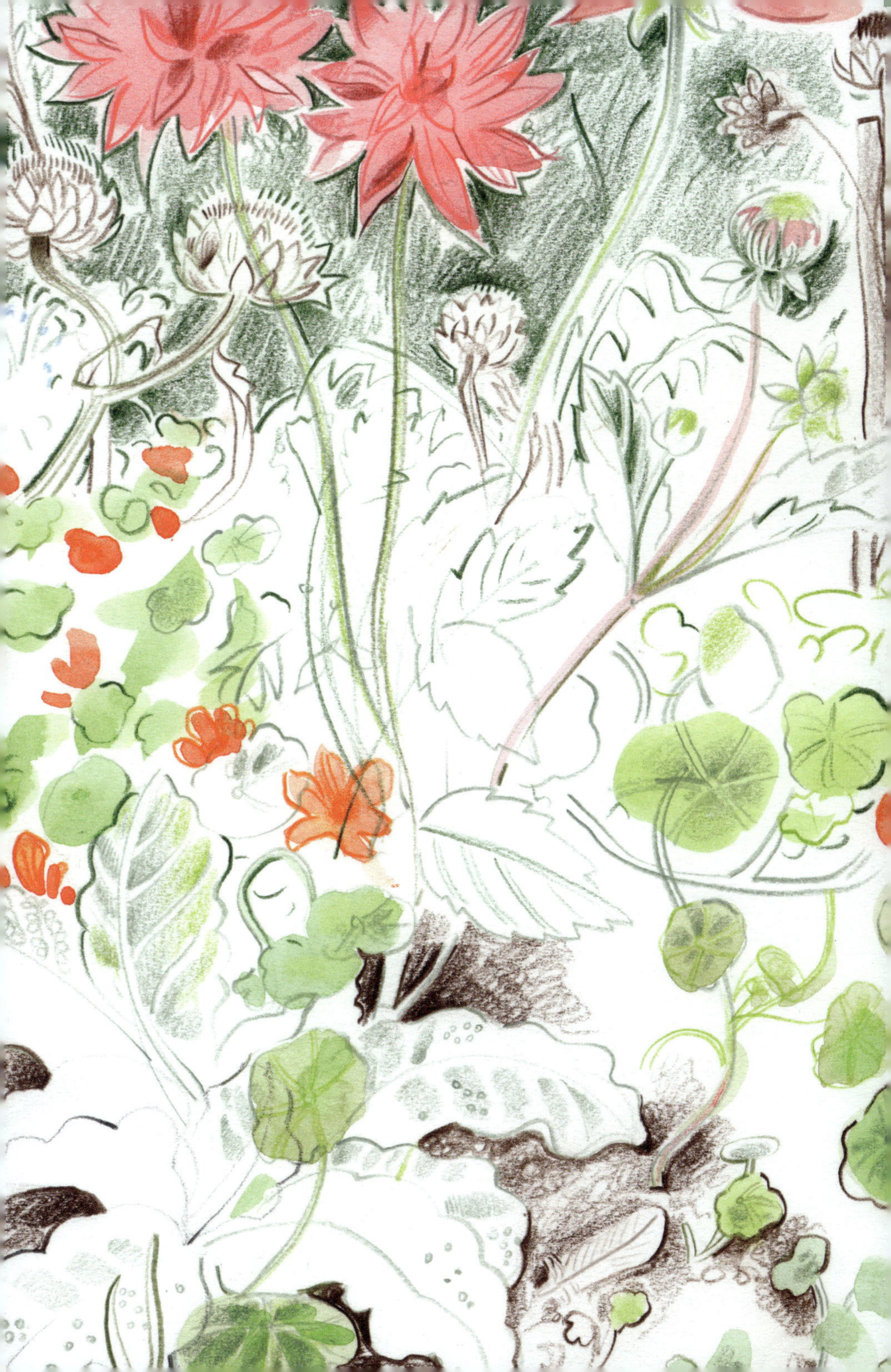

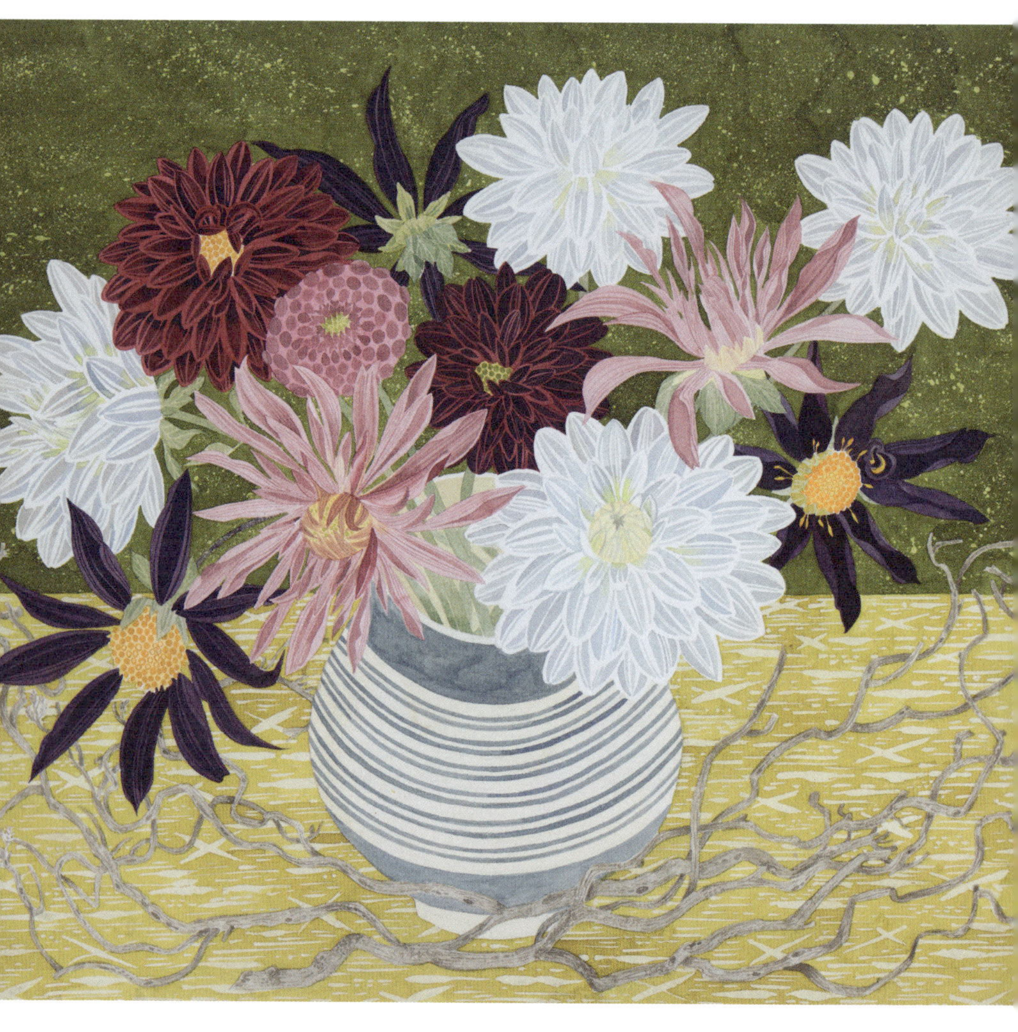

Dahlias in a Striped Cup
Watercolour, 2022

Page 16: *Dahlia sketch*
Watercolour and pencil, 2019

INTRODUCTION

THE WORLD IN YOUR GARDEN

CHRISTOPHER STOCKS

My garden may be small, but it contains almost the entire world. Looking out from my kitchen, I can see plants from Chile, Ethiopia, Korea, Turkey, Greece and Georgia, Portugal and Peru, Majorca and Mexico, China and the USA, not to mention Corsica and Japan, Serbia, Siberia, South Africa and the Canary Islands. This astonishing range makes it sound as though my garden is packed with rare and exotic specimens, but you might be surprised to learn how many of the flowers we take for granted come from the far corners of the globe. Nasturtiums originate in the mountains of Peru. Pelargoniums come from South Africa, as do most species of gladiolus and agapanthus. Camellias hail from China and Japan, as do daylilies, Japanese anemones, and the ancestors of many of our finest roses and peonies. Busy lizzies can be found growing wild in Kenya and Mozambique, while dahlias come from the forests of Guatemala and Mexico.

These, and thousands of other popular plants, have found their way into our gardens over hundreds (and in some cases thousands) of years, carried by crusaders, traders, priests, pilgrims, prospectors, diplomats, soldiers, scientists, botanists and, more recently, amateur and professional plant-hunters. Many of them have settled in as unmodified natural species, from the Turkish *Cyclamen coum* to the Tunisian *Iris unguicularis*, and from the giant *Echium pininana* from La Palma in the Canary Islands to *Geranium maderense* from Madeira. Others, though, have gradually changed – sometimes beyond all recognition – thanks to the attentions of farmers and horticulturalists, amateur and professional, who have repeatedly selected, grown and hybridized plants with unusual colours or leaves, for decorative or culinary purposes.

Artichokes offer an excellent example of this process. There's no such thing as a wild artichoke, but they appear to have been created artificially during the centuries of the Roman empire, when farmers started taking wild cardoons and saving only those with larger and larger flowers. Dahlias are a much more recent example. The first wild species were brought back from Mexico to Spain in the late eighteenth century, and they began arriving in Britain around 1804. Their vivid colours caused a sensation, and within thirty years the entire country was gripped by dahlia fever. By happy chance they proved easy to grow and to hybridize, and by the end of the nineteenth century the three or four original species had been transformed into an extraordinary ten thousand different cultivars, with a plethora of different flower shapes in virtually every colour under the sun.

Yet dahlias can also teach us another lesson: the vulnerability of artificial cultivars. One of the definitions of a natural species is that it comes true from seed – that is, if you grow seeds from a plant of, say, *Cyclamen coum*, you'll get offspring that (with rare, hybrid or mutated exceptions) look pretty much identical to their parents. That's rarely the case with artificial hybrids, whose genetic parentage is often extremely mixed. When I wrote my first book, *Forgotten Fruits*, I remember finding a fascinating passage in Charles Darwin's *The Variation of Plants and Animals Under Domestication*, first published in 1868. In an experiment he conducted in the gardens of Down House, his home in Kent, he explains how 'I raised 233 seedlings from cabbages of different kinds, which had purposely been planted near each other, and of the seedlings no less than 155 were plainly deteriorated and mongrelised; nor were the remaining 78 all perfectly true.' As I commented on cabbages at the time, 'That we now have so many different types – drum-headed or flat; round, oval, pointed or conical; winter, spring, summer, autumn; red, white, green; savoy – is a testament to generation after generation of dedicated gardeners. Few of us give cabbages a second thought, but it is actually a minor miracle that a handful of old varieties have survived at all.'

What's true of cabbages is even more true of dahlias. When they fell from fashion in the early twentieth century, few nurserymen bothered to maintain their stock of increasingly unsaleable cultivars, and, one after another, the Victorian varieties died out. Although many new varieties have been bred since then, of the ten thousand-odd cultivars available to Queen Victoria, just three lone examples remain.

Researching and writing *The Book of Garden Flowers* has been a captivating adventure, and surprisingly different from researching my previous collaboration with Angie Lewin, *The Book of Wild Flowers*. British wild flowers have a history that stretches back, in most cases, at least ten thousand years, to the end of the last Ice Age. They are embedded in our folklore and our culture as well as the wider ecology of the countryside, whether in the intimate relationship between devil's-bit scabious and marsh fritillary butterflies or the spread of buck's-horn plantains since main roads began being salted against icy conditions in the 1960s.

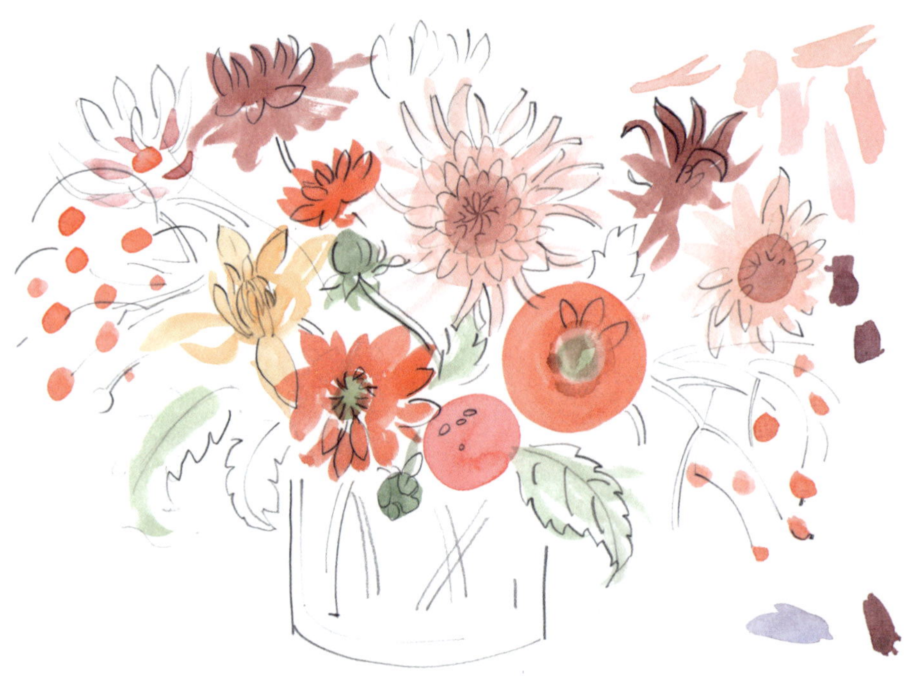

The flowers we grow in our gardens may all originally have derived from wild species, often from far-flung corners of the world, but their history on these islands can usually be measured in (at most) a few hundred years. And, while we have no idea who introduced snowdrops to the British Isles, or when they arrived, we often know exactly when particular garden flowers made their first appearance on the scene. More than that, we also know who brought them here, and who it was who developed a specific cultivar. This makes *The Book of Garden Flowers* as much a social study as a horticultural history, and it is this that has made it so rewarding to research. I loved discovering the stories about such people as Lewis 'Luly' Palmer, who developed hardy agapanthus cultivars for the British climate at Headborne Worthy near Winchester, or Ellen Willmott and her ghostly sea holly – a tale that turned out to be not quite what it seemed. Or how about Helen Ballard, whose tireless work in the shadow of the Malvern Hills led to many of our finest modern hellebores? Or Margery Fish and her Somerset astrantias?

It has been fascinating to dig and delve into these stories as I've researched this book. Some of them are well known, while others have rarely been told before – and one or two remain tantalizingly mysterious. The various stories behind the cultivated varieties of so-called poppy anemones, *Anemone coronaria*, are fascinating, but despite my best efforts I've been unable to track down the origins of the intriguingly named 'Mr Fokker', apart from establishing that it was first sold by a Dutch company in 1928. For the rest, I hope that you enjoy reading about them as much as I have enjoyed discovering their history.

Christopher Stocks, Isle of Portland, Dorset

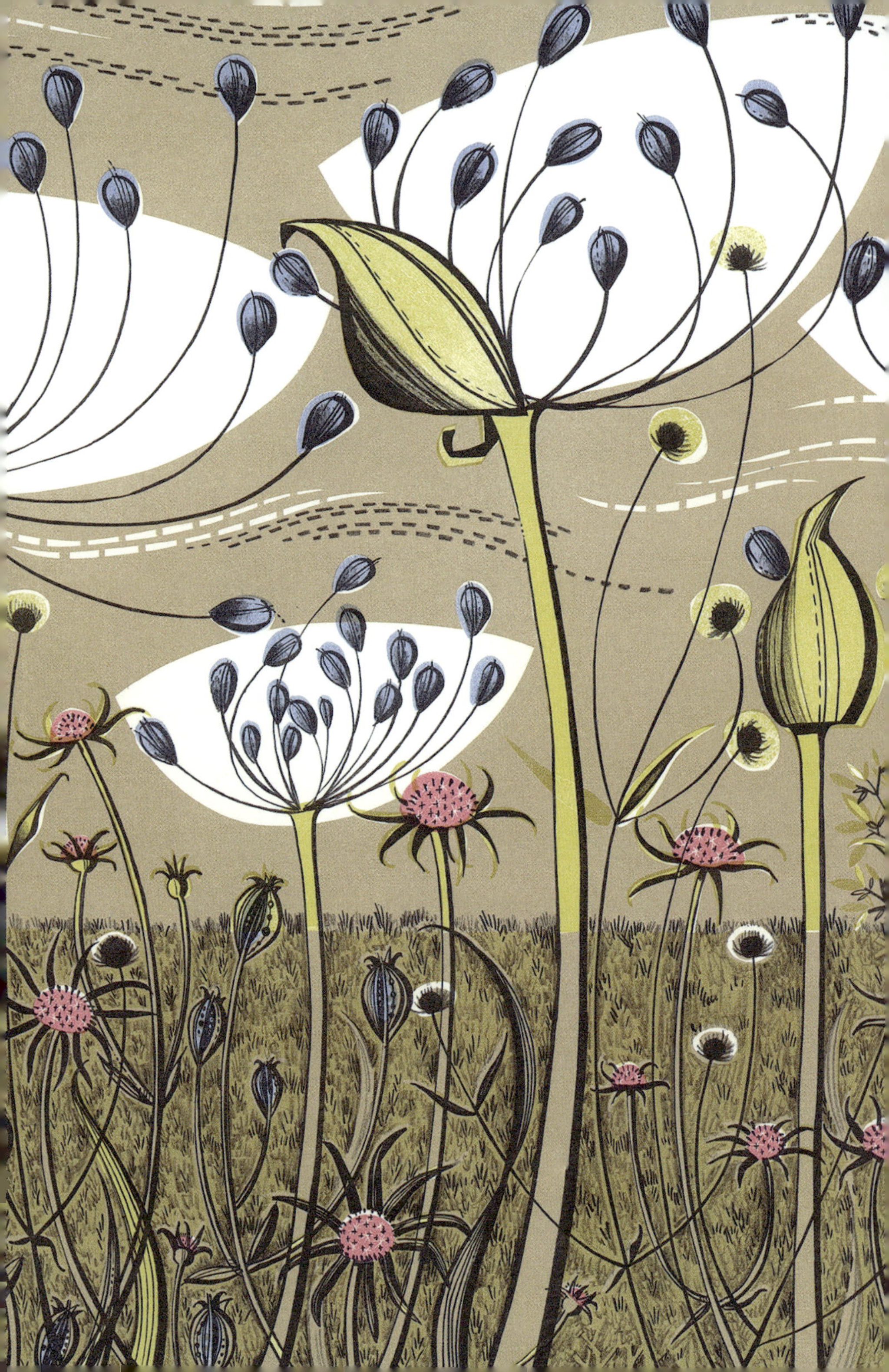

AGAPANTHUS

Agapanthus (or agaphantoms, as my Spanish-born friend Azucena sometimes refers to them) make a striking addition to any garden, with their strappy leaves and bursts of blue or white flowers on long stems. They belong to a small family of subtropical plants, most of which are native to South Africa, with a few other species found in Lesotho, Mozambique and Swaziland. The first species to arrive in Britain, some time in the early seventeenth century, was *Agapanthus africanus*, shipped back from the Cape of Good Hope by traders of the Dutch East India Company. These blue, trumpet-shaped flowers were unlike anything else growing in European gardens at the time, and must have caused a sensation, although they are tender and would have grown best under glass. In the wild *A. africanus* is found only in Western Cape province, where it grows

Agapanthus (detail)
Lithograph, 2008

on sandy soils from sea level up to a thousand metres or so, growing around 70 cm high and flowering between December and February, during the rainy season. Its main pollinators are bees and sunbirds, and it flowers most profusely after summer fires have cleared the ground around the plants – not the easiest conditions to re-create in the average garden. Most of the plants sold as *A. africanus* today are actually hybrids, and much easier to grow than the true species, although they still need protection from frost.

Luckily a second, hardier species was discovered later in the seventeenth century, and this is the plant behind almost all garden agapanthus today. *Agapanthus praecox* ('praecox' meaning early-flowering) is larger than *A. africanus*, growing to a metre tall, with mid-blue or occasionally pure white flowers. It has three subspecies: *praecox*, which grows in the Eastern Cape and has arching leaves like our garden agapanthus; *orientalis*, which is also found in KwaZulu-Natal; and *minimus* from Western and Eastern Cape – which, as its name suggests, is smaller in size, and is the species behind most dwarf garden cultivars. *Agapanthus praecox* can be grown outdoors in the warmest parts of southern Britain, but it was often killed by winter frosts, which limited its popularity further north – at least until the 1940s, when the Hon. Lewis Palmer decided to do something about it.

Lewis Palmer (better known as 'Luly' to his friends) is one of those engaging characters that make British horticultural history so rewarding. Born in 1894, he was the youngest son of the second Earl of Selborne, who ran Southern Africa for the British government

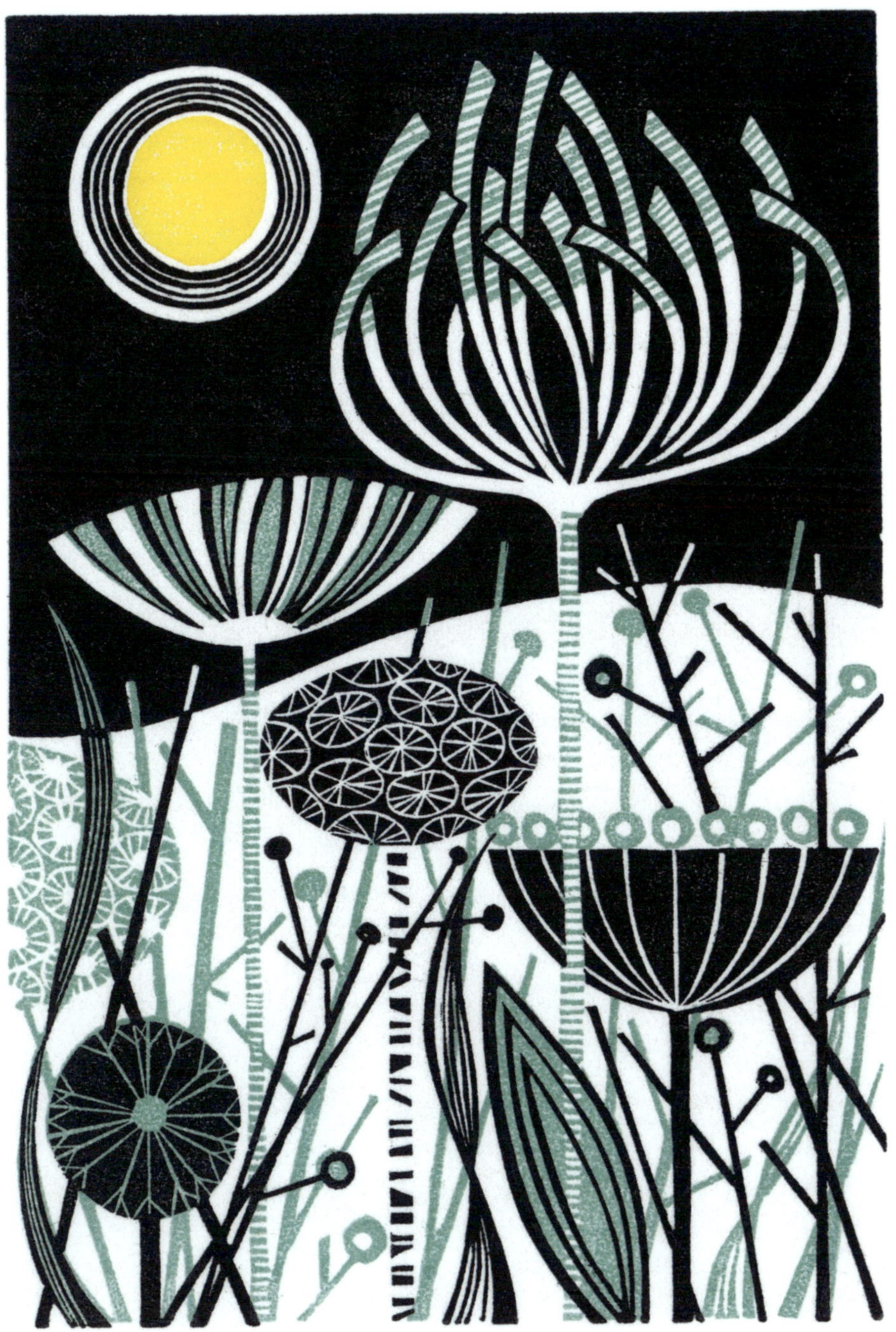

for a while. He went to school at Winchester, fought in the First World War and went on to Oxford, which he followed with a very gentlemanly career as a director of various City companies and treasurer of the Royal Horticultural Society, whose council he sat on for thirty years. In 1922 he married Dorothy Loder, who came from gardening royalty: her father, Gerald Loder, had bought Wakehurst Place in Sussex in 1903 and spent the next three decades creating one of the most celebrated gardens in England. Today they are owned by the National Trust and run by the Royal Botanic Gardens, Kew.

Palmer gardened in London and Hampshire, first at Wonston and later at Headbourne Worthy Grange, just north of Winchester, which had thin chalk soil and was, he claimed, the coldest garden in the country. As such it was the perfect place for his experiments with agapanthus seedlings. Christine Skelmersdale, owner of Broadleigh Bulbs and one of the leading horticulturists of our own time, visited 'Uncle Luly' a year or so before he died, and believes that, 'As far as I know the evolution of Luly's agapanthus was entirely serendipitous. At that time, 1970, the ones grown here were entirely evergreen and either grown in Cornwall or kept in a glasshouse. They were certainly not popular.' She recalls that Palmer's research into what became his famous 'Headbourne hybrids' started with a packet of mixed seed he received from the famous Kirstenbosch botanic gardens in Cape Town. Agapanthus are naturally variable and hybridize easily between different species, but there must have been years of painstaking experiment before Luly came up with cultivars that looked good but also – crucially – could withstand British winters.

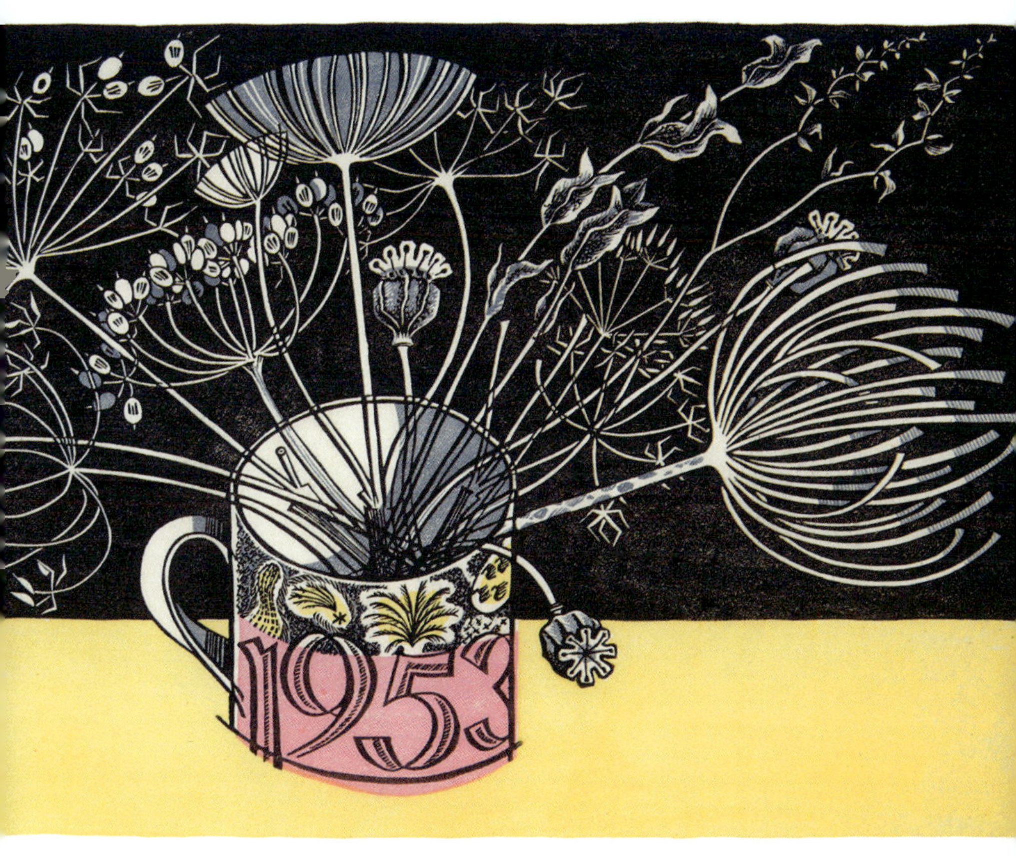

Palmer's obituary, published in the *Journal of the Royal Horticultural Society* in 1971, goes into a little more detail, explaining how the Headbourne hybrids 'emerged out of a thorough study of a much confused genus during which he not only grew a large collection of them but also visited South Africa. There he studied all the plants he could find in their native haunts as well as checking all the references to them in the literature and working with South African botanists.' We have a lot to thank Uncle Luly for.

Coronation Mug
Linocut, 2003

Overleaf: *Agapanthus II*
Screenprint, 2013

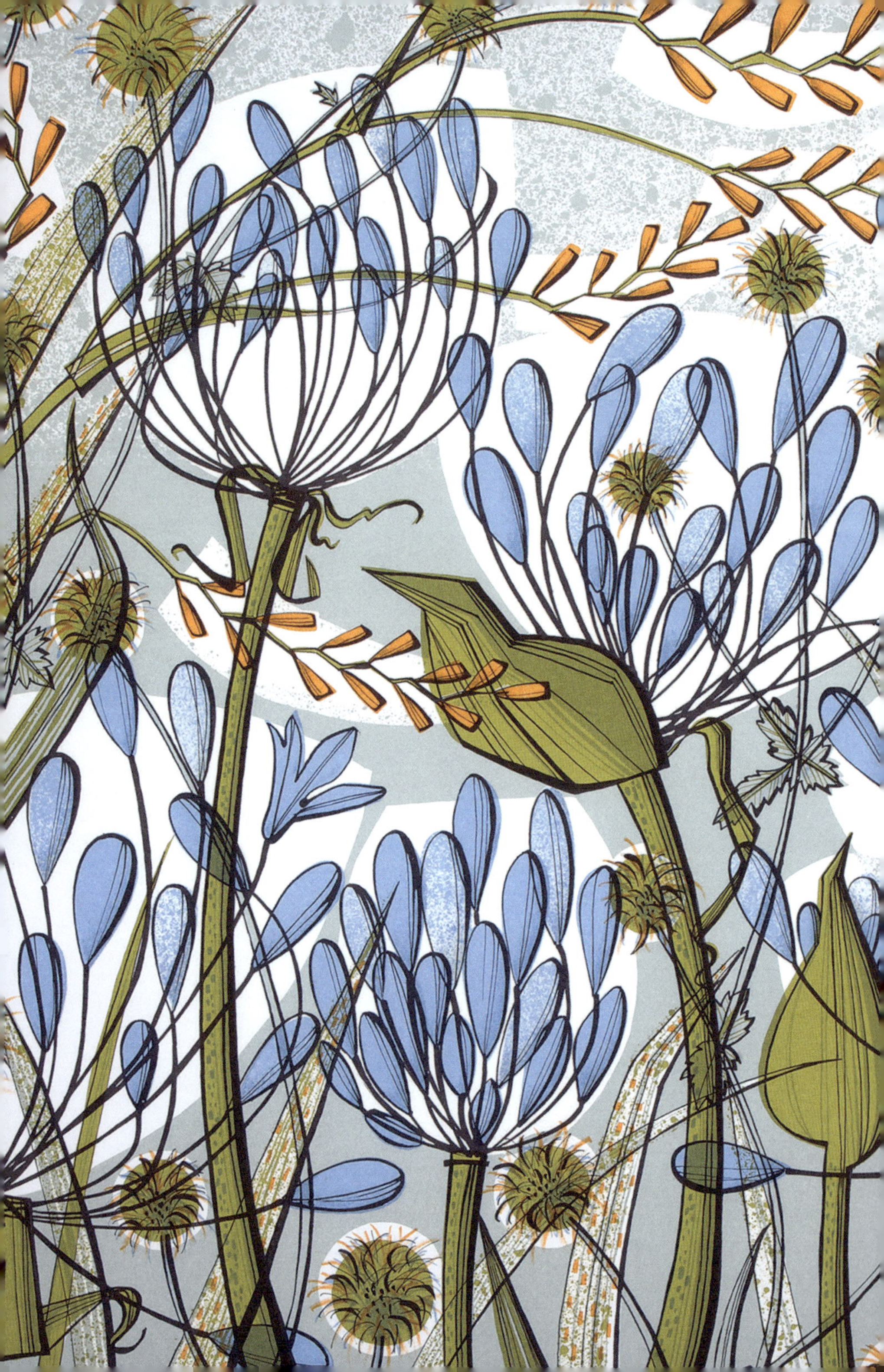

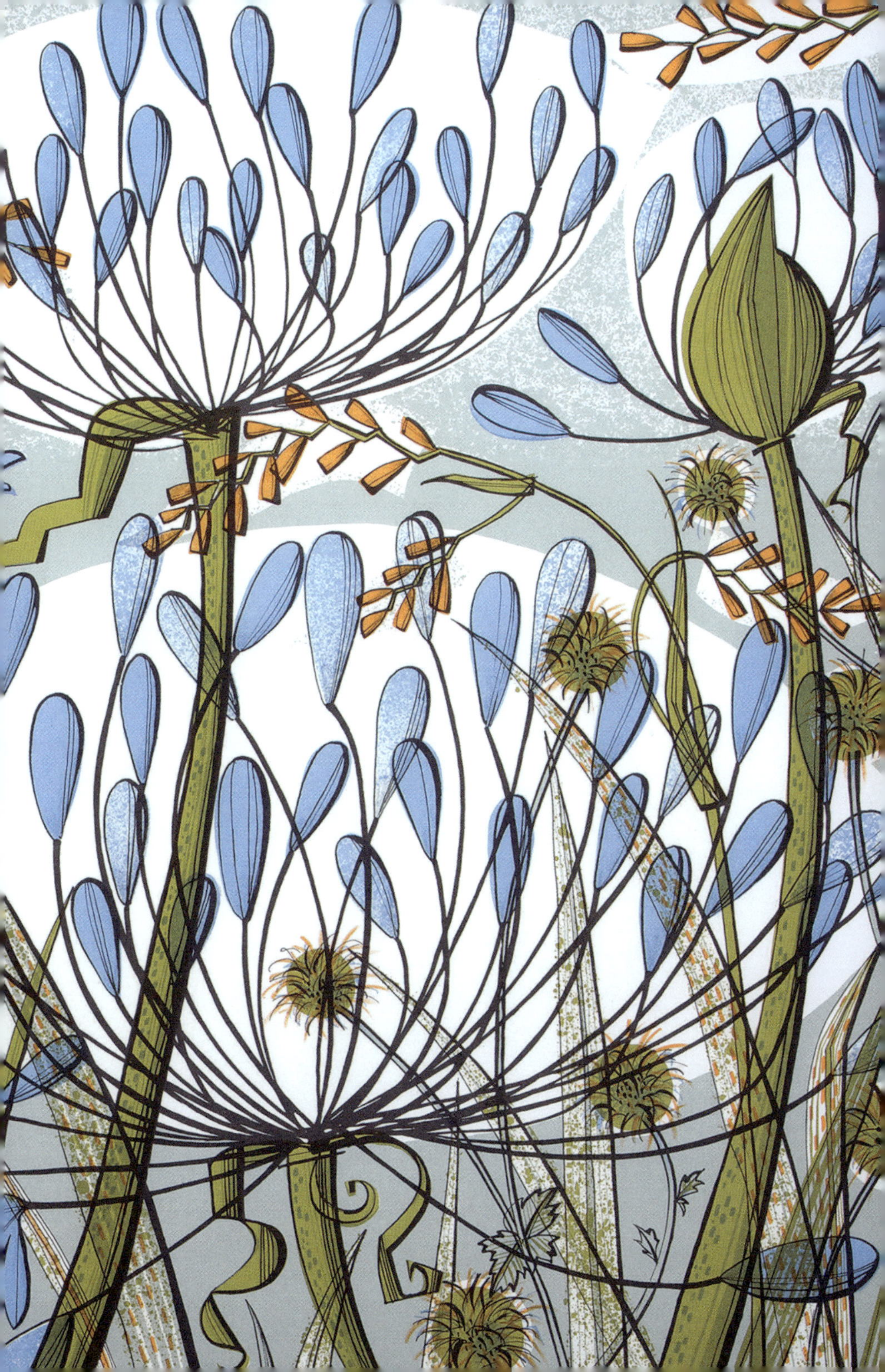

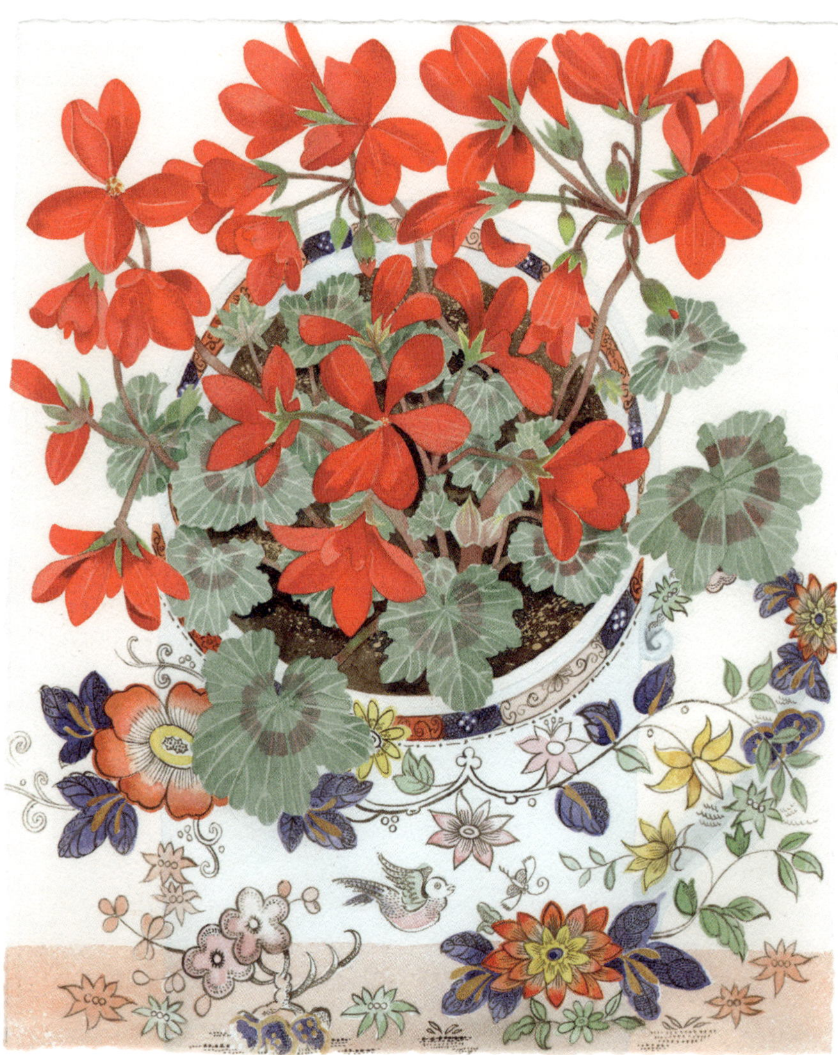

2

PELARGONIUMS

First, the name. Most of us call those red-flowered plants in window boxes geraniums, even when we know they're really called pelargoniums. Perhaps it's simply because the word 'pelargonium', although more botanically accurate, has an extra syllable and is less elegant than the word 'geranium', but whatever the reason, the muddle goes back a long way. There are around 280 species of pelargonium, scattered across Asia, Australasia and South Africa, which is where they were first encountered by European travellers in the late seventeenth century. The German-born botanist Johann Jakob Dillenius suggested calling them pelargoniums in 1732, but he was overruled by his far more famous friend, Carl Linnaeus, who insisted on naming them geraniums, although they're only distantly related to European geraniums, such as cranesbill (also known as

Pelargonium in Floral Cup
Watercolour, 2024

herb robert). They were eventually renamed pelargoniums in 1774, but the confusion has persisted.

Most garden pelargoniums are hybrids derived from a relatively small number of South African species. One of the first to arrive in Europe was *Pelargonium peltatum*, the so-called ivy-leaved geranium, which gave us those long, trailing varieties of pelargonium that look so good cascading from pub window boxes. Possibly the first person to grow it in Britain was Mary Somerset, Duchess of Beaufort, who created remarkable gardens at Badminton House in Gloucestershire, and at Beaufort House by the River Thames in Chelsea (on the site of present-day Beaufort Street). Buoyed by a ducal income, she amassed vast numbers of exotic plants, grown in then-rare – and enormously expensive – glasshouses, and *P. peltatum* was recorded in her collection in 1701.

Although several other species were introduced around the same time, including the rose-scented geranium, *Pelargonium capitatum*; *P. zonale* with its decorative leaves (ancestor of our zonal pelargoniums); and the red-flowered *P. inquinans*, it wasn't until 1844 that a cunning cultivator crossed the last two species and came up with a sturdy, scarlet-flowered hybrid. Named 'Tom Thumb' after the famously diminutive American teenager (real name Charles Stratton) who toured Britain that year with the impresario P. T. Barnum, it quickly established itself as the period's most popular pelargonium. Before long 'Tom Thumb' was being bedded out in vast displays, and it caught the eye of none other than Charles Dickens, who became mildly obsessed with the cultivar.

At Gad's Hill Place, his country house in Kent, Dickens had well over a hundred pots of scarlet pelargoniums, ranged on 'geranium theatres' below the two bay windows at the front, as well as bedded out in the garden and conservatory. Their flowers brightened the dining table, and it's said he even wore them as a buttonhole. Dickens's passion for pelargoniums exceeded his notorious love of mirrors, and after he had the drawing room lined with them (mirrors, that is, rather than pelargoniums), his younger daughter, Kate, said to him, 'I believe papa, that when you become an angel your wings will be made of looking-glass and your crown of scarlet geraniums.' Whether she got her wish or not, Dickens died at Gad's Hill on 9 June 1870, at the age of just 58, and his pelargoniums outlasted him for only a couple of months, forming part of a sale that took place on 10 August the same year.

Overleaf: An engraved
wood block for *Pelargonium*

Page 31: *Pelargonium*
Wood engraving, 2024

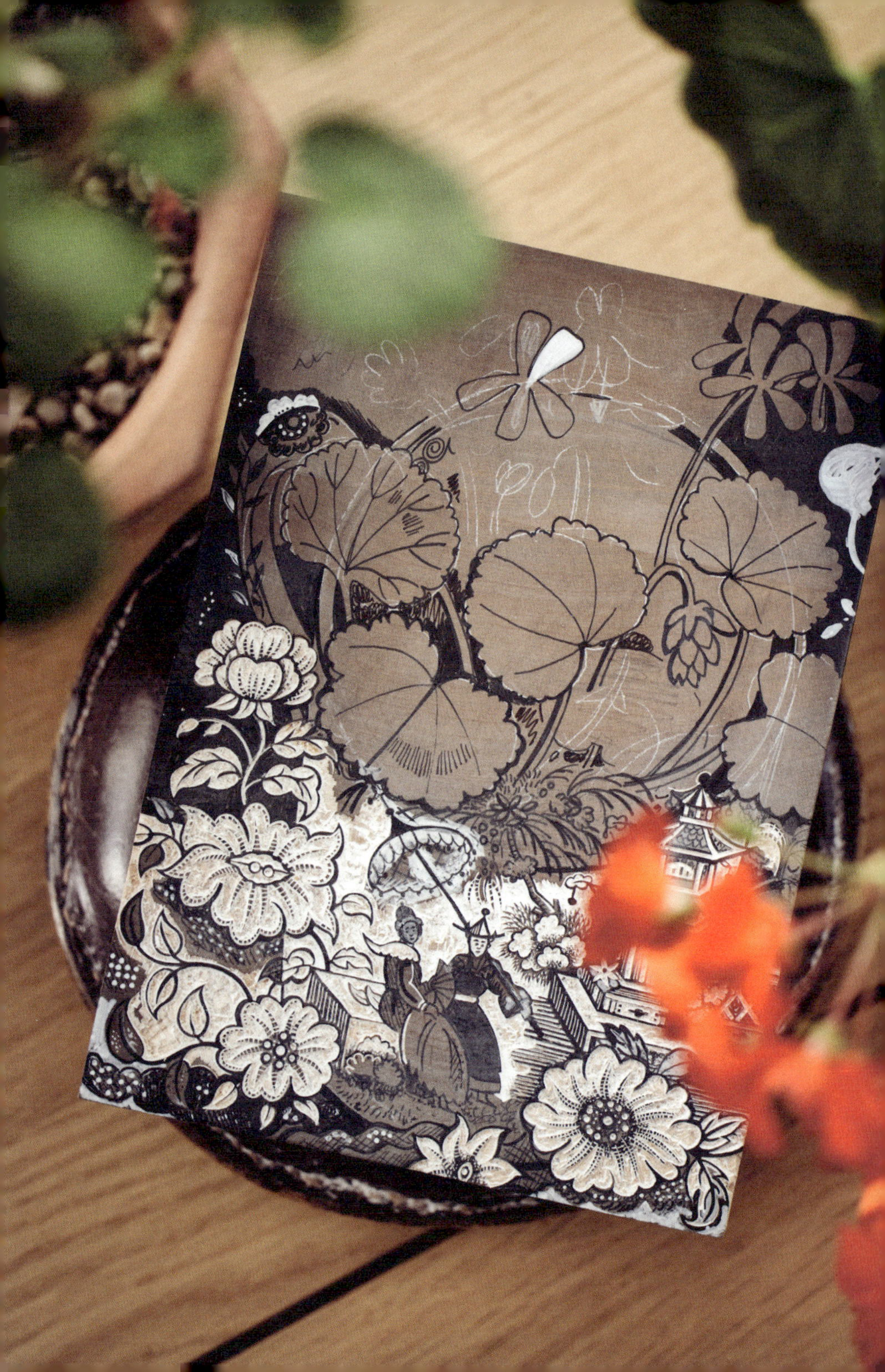

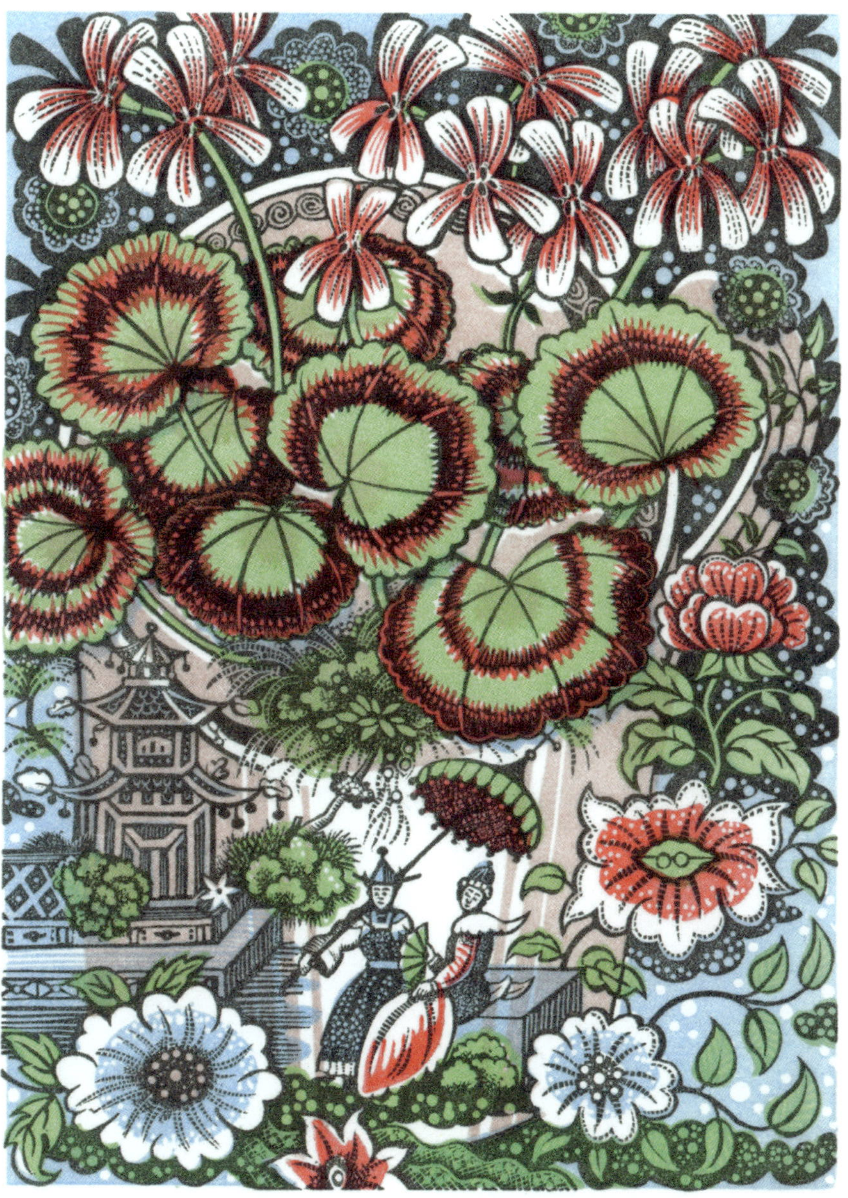

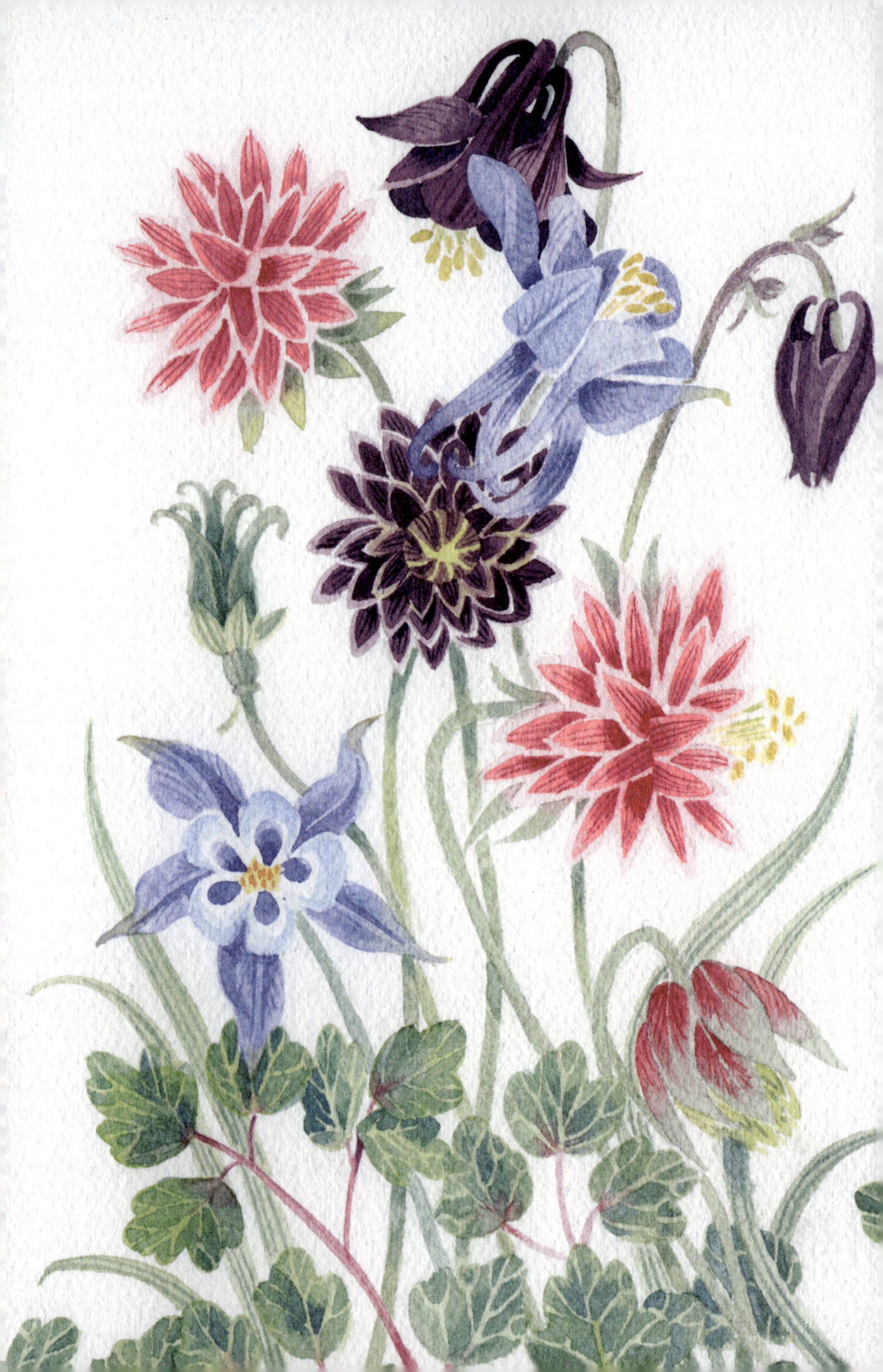

3

AQUILEGIAS

If there are a few plants that no cottage garden should be without, surely aquilegias have to be among them. Granny's bonnets or columbines, as they're often called, are some of the easiest and most undemanding flowers to grow. They're also striking, pretty and remarkably varied, although sometimes a martyr to mildew. There are more than sixty species in the aquilegia family, but the most commonly cultivated is *Aquilegia vulgaris*, which grows right across Europe (with the exception of Scandinavia), from Britain as far east as Russia. In the wild they come in shades of purple, but over the centuries many hybrids and selected forms have been developed, introducing a wide range of colours and double flowers.

More recently, some of the other species have been introduced into cultivation, including the beautiful yellow *A. chrysantha* from the USA, the delicate yellow-and-purple *A. oxysepala* from Japan, the pineapple-scented *A. fragrans* from Pakistan and the sky-blue *A. coelurea* from the Rocky Mountains. Still, there's more than enough variation among common-or-garden *A. vulgaris* cultivars to satisfy most gardeners.

One of the most popular is the starry red-and-white double-flowered 'Nora Barlow', which sounds as though it was named after a sitcom character but actually has a far more distinguished pedigree. Although similar forms have been around for centuries, 'Nora Barlow' was introduced in the 1960s by Alan Bloom, a remarkable horticulturist who, in a long life (he died at the age of 98), was responsible for hundreds of other plant cultivars, as well as promoting those amoeba-shaped beds filled with heathers and dwarf conifers that graced many a 1970s garden. Since its launch, several other variants have been added to what is now referred to as the Barlow Series, including the self-explanatory 'Black Barlow', 'Bordeaux Barlow', 'Pink Barlow' and 'Barlow Blue'.

Bloom called his cultivar 'Nora Barlow' after the person who had originally given him the seeds. Nora Barlow may not sound like a name to conjure with, but once you learn that her maiden name was Darwin, things start to become more interesting. In fact she was Charles Darwin's granddaughter, and followed in his footsteps by training as a botanist and geneticist. She married the civil servant Sir Alan Barlow in 1911 and had six children; one of their grand-children is the poet Ruth Padel.

Among Barlow's other achievements was editing her grandfather's diaries, as well as a lifelong study of plant genetics. She wrote learned papers on primroses but also had an abiding interest in aquilegias, which are promiscuous interbreeders, and set a lot of seed. Apparently she grew many varieties in her own garden at Boswells, on Hogtrough Lane near Wendover in Buckinghamshire, and it was presumably some of that seed she gave to Alan Bloom. Barlow died in 1985 at the ripe old age of 103, but thanks to Bloom her double aquilegia lives on.

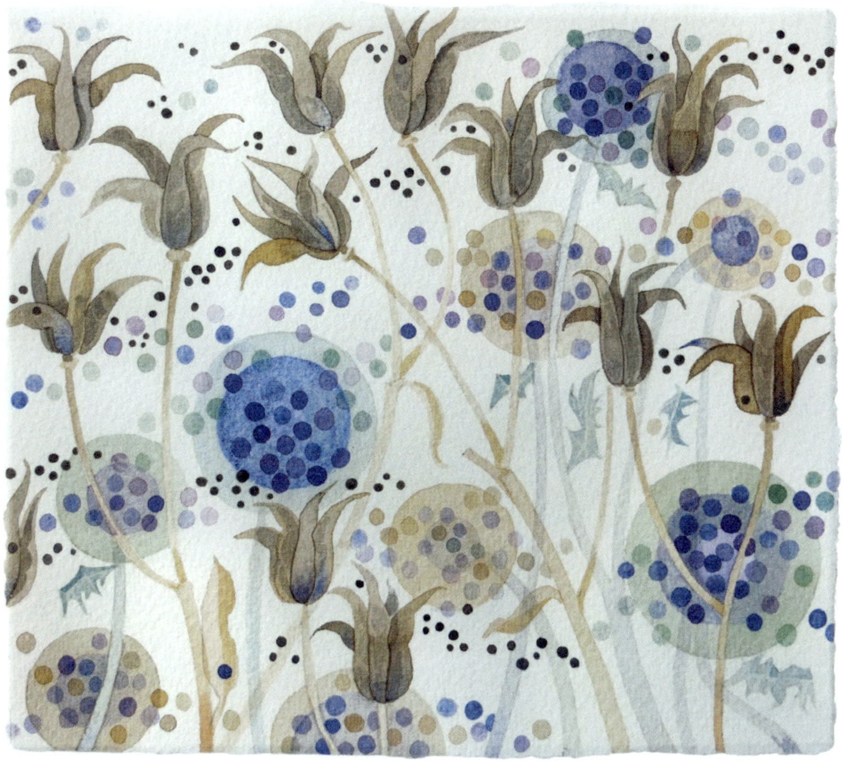

Echinops and Aquilegia, watercolour, 2020
Overleaf: *Honesty, Aquilegia and Artichokes*, watercolour, pen and ink, 2020

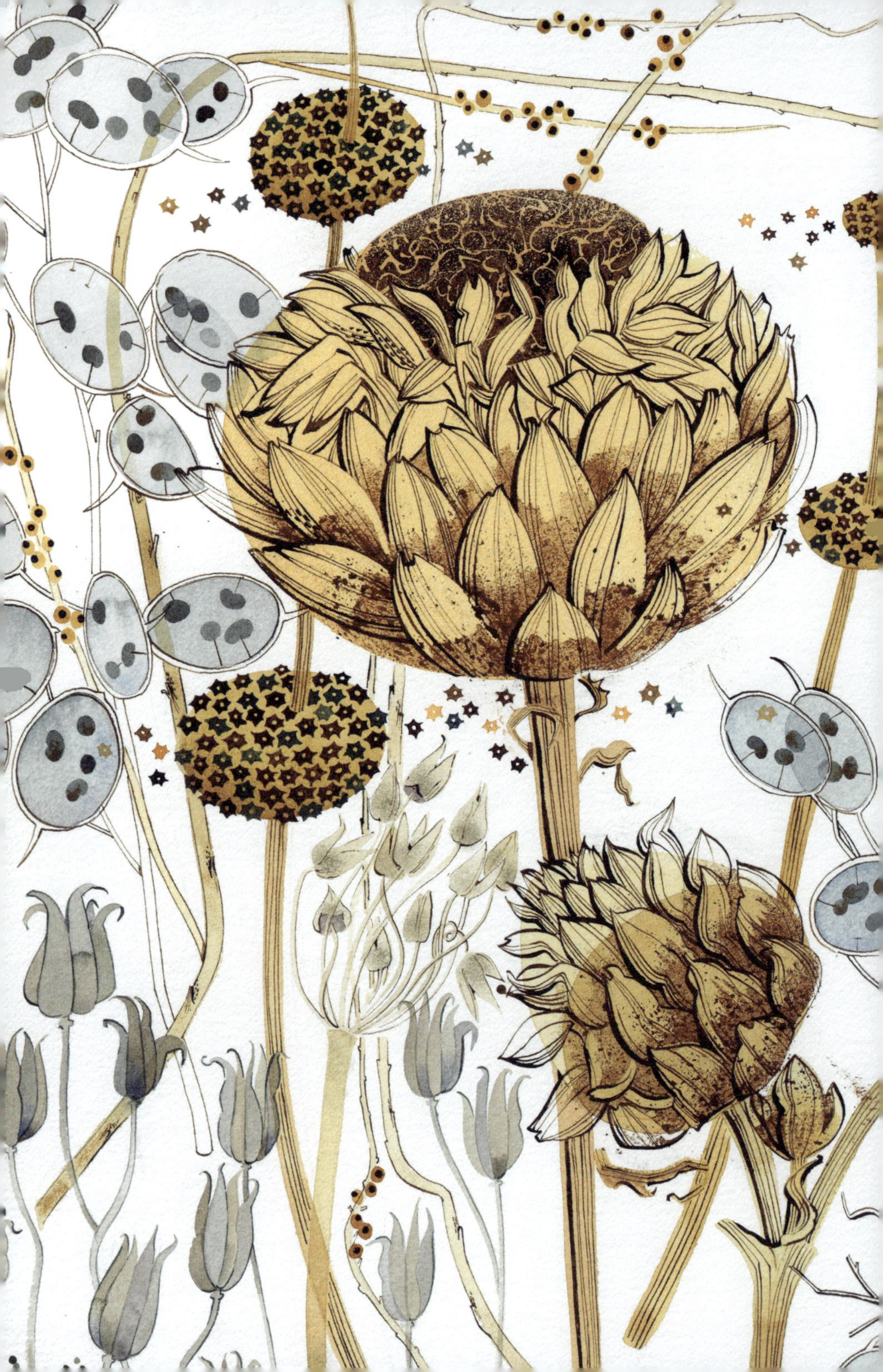

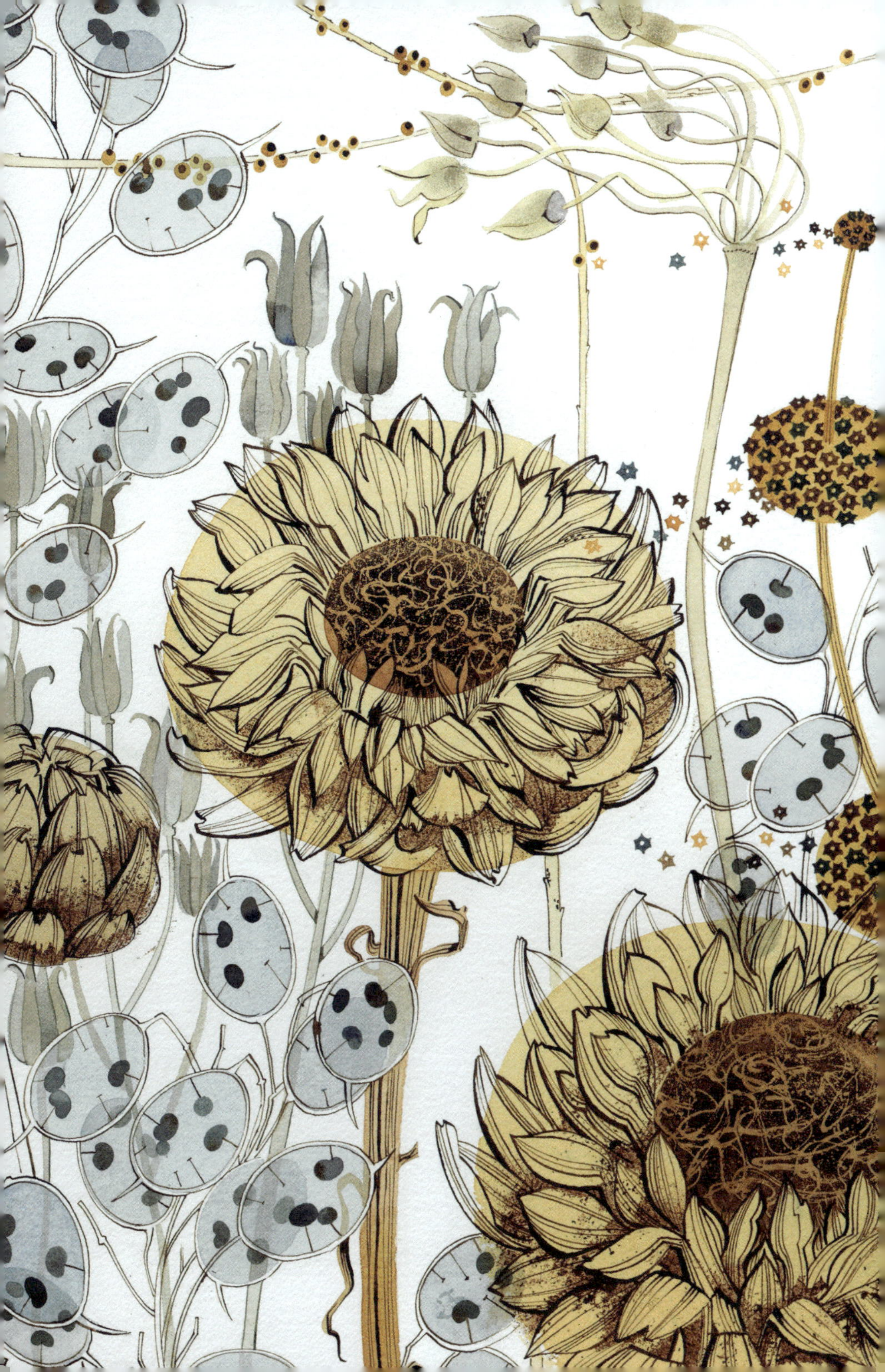

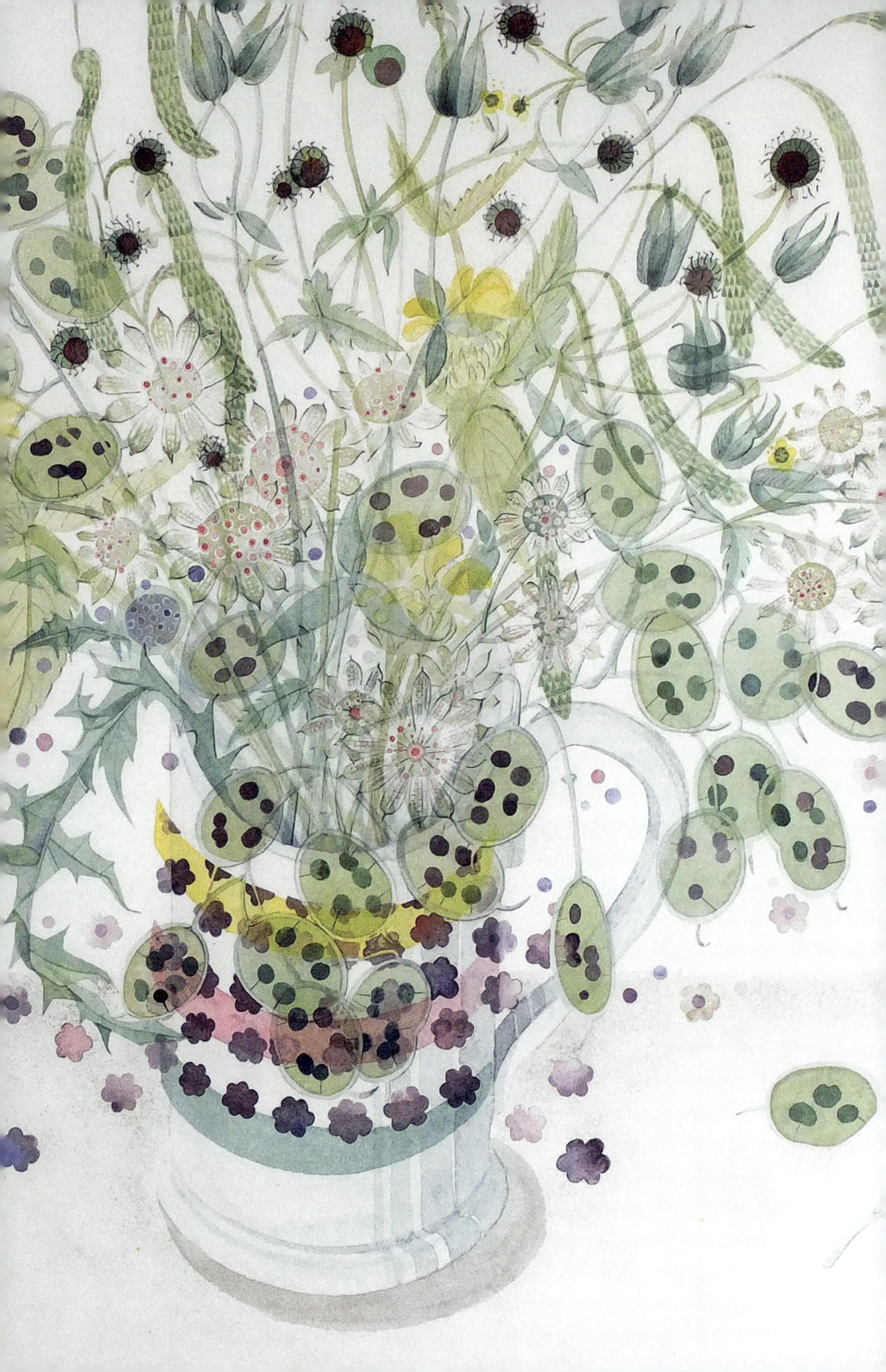

4

HONESTY

I didn't think about my parents' garden at the time, but I suppose you would have called it a woodland garden. Although the area closest to our house caught the sun, much of it was lightly shaded by full-grown trees – wych elm and ash, and a magnificent beech, from whose smooth grey branches hung our swing. The wood ran down to a rocky stream, and, although it lacked bluebells or wild garlic, was carpeted with cow parsley in the spring. In the topmost branches lived a raucous colony of rooks, who during nesting time gave a great communal shout whenever we went outdoors, and dropped broken twigs on the flowerbeds. My parents grew lots of plants that thrived in semi-shade, such as lungworts and hellebores, but the one I liked most as a child was honesty, with its sensuously smooth, pearly seed discs – pale moons that you could rub apart in your hands, revealing the flat brown seeds inside.

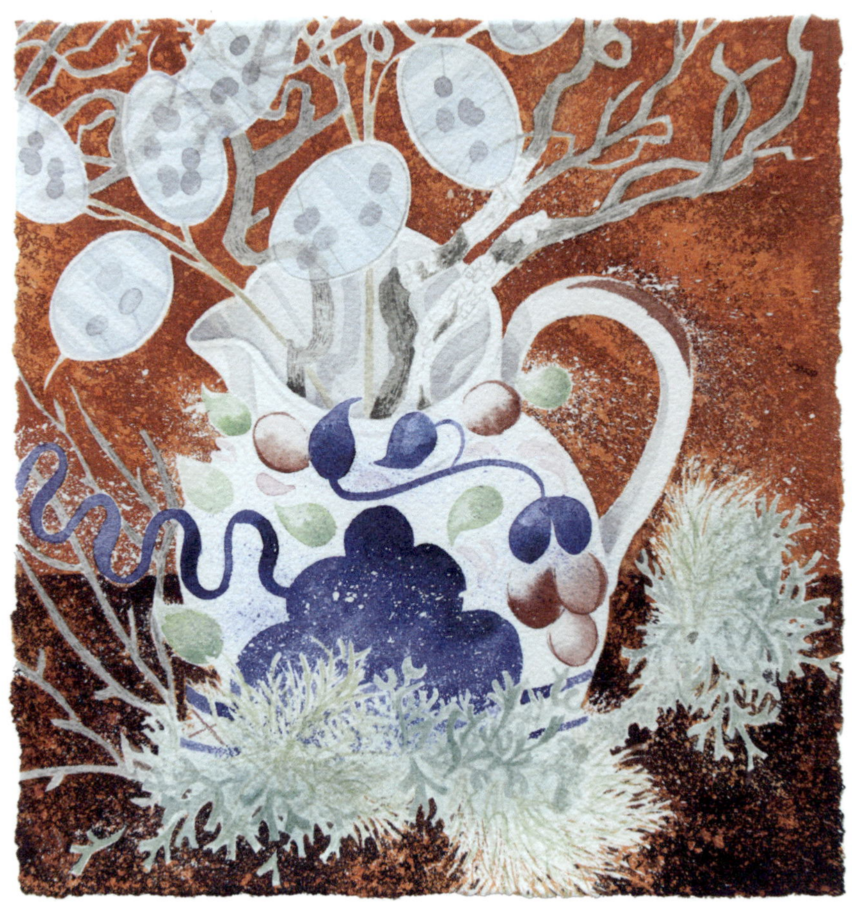

Lunaria annua, to give honesty its botanical name, grows on woodland edges all over Europe, and while it isn't a British native, it has been growing here for many centuries. Like wallflowers, aubretia and night-scented stocks, honesty is a member of the cabbage family, the Brassicaceae, and has the simple four-petalled flowers and sweet smell that are shared with many of its relatives. Despite its Latin name of 'annua', it's actually a biennial, flowering only in the second year before dying off, but, like most short-lived flowers, it readily sets seed; the 'Lunaria' part refers to its moon-like seed discs.

Honesty was first recorded in Britain in 1597 in John Gerard's *Herball*, along with a second species, the perennial, white-flowered *L. rediviva*, which has a similar distribution across Europe. *L. annua* was a familar garden plant in Gerard's day, and already known to herbalists as Lunaria, although Gerard noted that 'among our women it is called Honestie'. He lists several other engaging common names – bolbonac or satin flower, penny flower or money flower, silver plate or pricksong wort – and notes that while both species 'are set and sowne in gardens…the first hath bin found wild in the woods about Pinner and Harrow on the hill 12 miles from London, and in Essex likewise about Horn-church', while 'The second groweth about Watford, fifteene miles from London.'

Honesty is as popular now as it was in Gerard's time, and there are various cultivated forms, including the variegated white-flowered *L. annua var. albiflora* 'Alba Variegata', which comes with an Award of Garden Merit from the Royal Horticultural Society. They're easy to grow and good for attracting pollinators, although, like other brassicas, they are also popular with the caterpillars of cabbage-white butterflies, which sometimes feast on their leaves before moving on to your nasturtiums.

Gaudy Welsh Jug with Honesty and Lichen
Watercolour, 2020

Page 38: *Summer Garden, Edinburgh*
Watercolour, 2018

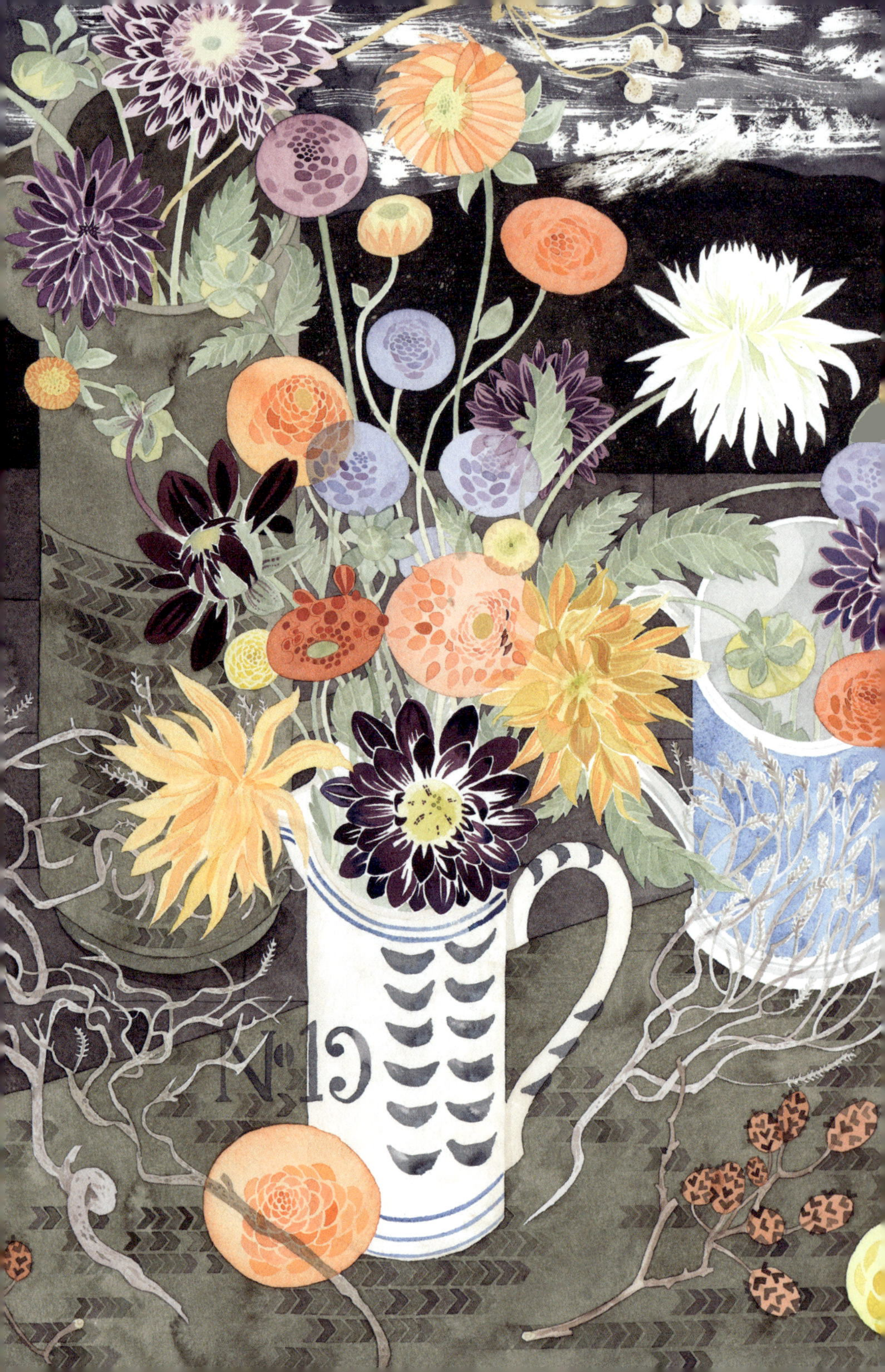

N° 19

5

DAHLIAS

There's something doolally about dahlias. They're big. They're blowsy. They come in an eye-jangling range of clashing colours, and vary in shape from simple single flowers to huge, ungainly blooms of the mop-top or bath-hat type. Their foliage is lush and sappy and beloved of slugs, and they drive dahlia-lovers equally wild with desire. In the autumn you can see entire front gardens devoted to dahlias, their livid 1970s disco shades impossible to ignore. There are getting on for 18,000 RHS-registered varieties to choose from, in every colour except blue, organized into fifteen groups based on the shape of their flowers, from Semi-Cactus to Double Orchid, Paeony to Pompon. Even their names are bonkers, from *Dahlia* 'Doris Bacon' to the Trumpian 'Orange Mullet', the titillating 'Top Totty' and the nasty-sounding 'Harvest Inflammation', not to mention the wilfully wacky 'Wicky Woo'.

The amazing thing is that all these thousands of varied flower forms and colour combinations derive from just a handful of wild species, which were introduced to European gardens as late as the nineteenth century. Their natural habitat is the high plains and forests of Mexico and Guatemala, where they were cultivated by the Aztec less as garden bedding and more as a root vegetable. Their tuberous roots are edible, although something of an acquired taste, being bitter and sharp and said to lurk somewhere between a radish and a potato.

Dahlias were first described by a botanist working for Philip II of Spain, Francisco Hernández, who spent six years in Mexico from 1571, although his book about his time there wasn't published until 1651, long after his death. The first three species were sent from Mexico to Spain in the late 1780s, and were grown by the botanist Antonio José Cavallines at the Royal Gardens in Madrid, flowering for the first time in 1789. It was Cavallines who gave them the name dahlia, to commemorate his Swedish colleague Anders Dahl, a pupil of Linnaeus; Dahl had died earlier the same year. Ever since, debate has raged over the correct pronunciation of the word, given its Scandinavian derivation; most of us say 'day-lia', but perhaps it really should be 'dar-lia'.

Although they were introduced to Europe as vegetables, their peculiar flavour never caught on. And that might have been curtains for the dahlia, except for a quirk of genetics. Like pansies, strawberries and sugar cane, dahlias are octoploid, with eight sets of homologous chromosomes, as well as a plethora of transposons – which in plain English means that they are naturally variable and, when grown

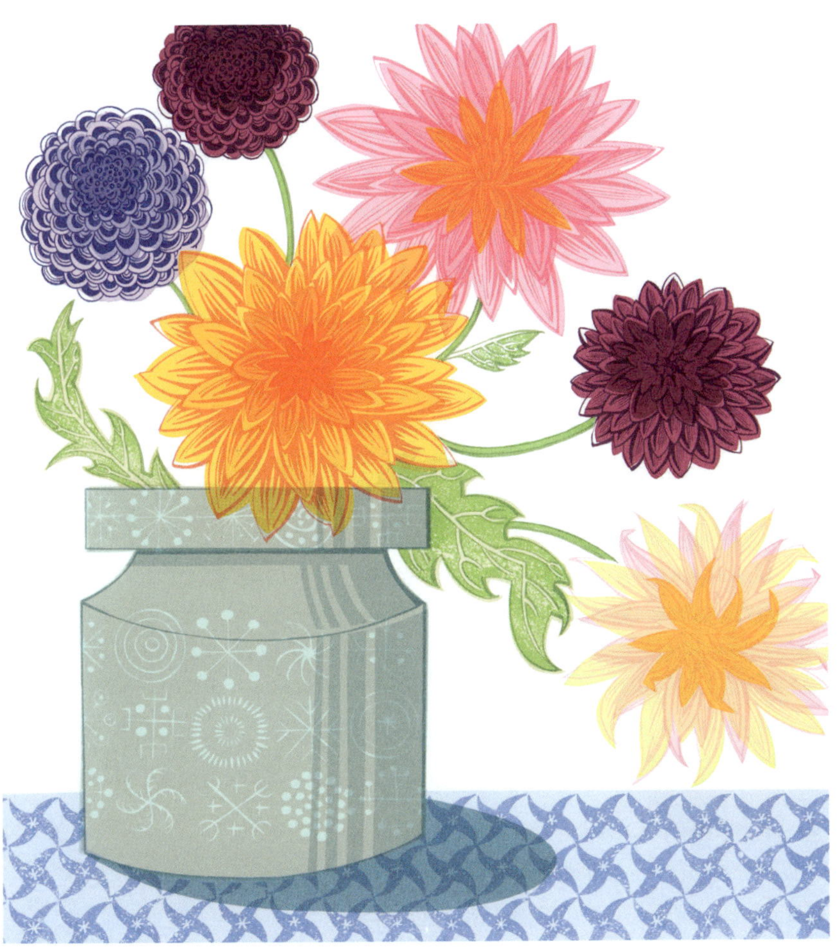

Dahlias and Totem
Screenprint, 2024

Page 42: *Dahlias with No. 19 Jug* (detail)
Watercolour, 2019

from seed, often generate new colours and flower shapes. Gardeners quickly noticed this attribute, and set to the work of hybridization.

It wasn't long before dahlias arrived in Britain, thanks to the vastly wealthy Welsh landowner Charlotte Hickman-Windsor, who accompanied her husband, the equally rich 1st Marquess of Bute, to Madrid while he was acting as ambassador to Spain. In 1798 she sent some seed from Madrid's Royal Gardens to the botanic garden at Kew; the resulting plants flowered once but then turned up their toes, presumably killed off by British frosts. They were introduced more successfully in 1804, thanks to the slave-plantation heiress-cum-political hostess Elizabeth Fox, Baroness Holland. Like the Marchioness of Bute, Lady Holland visited the Royal Gardens in Madrid, where Cavallines gave her some seeds of the still-novel dahlias he was growing there.

For some reason Lady Holland sent these back to her Italian librarian, Serafino Buonaiuti, at Holland House in London (now the setting for Opera Holland Park), and it was thanks to him that dahlias first became established in England. Buonaiuti is an intriguing figure. He was the opera librettist for the King's Theatre, contributed to the *Literary Gazette* and was the author of *Italian Scenery; Representing the Manners, Customs, and Amusements of the Different States of Italy*. In his spare time he worked for Lord Holland, designing one of the walled gardens at Holland House and looking after its fine library (which survived until 27 September 1940, when the house was gutted by fire during a German bombing raid). Why Lady Holland entrusted her librarian rather than her gardener with the precious dahlia seed isn't entirely clear, but Buonaiuti did at least manage to grow them.

According to Joseph Paxton, head gardener to the Duke of Devonshire at Chatsworth and later the designer of the Crystal Palace, Buonaiuti managed to raise a few plants from the seed, which produced purple or crimson flowers in the following season.

Paxton was a huge dahlia enthusiast himself, and indeed was the first to write about them in any detail. In his *Practical Treatise on the Cultivation of the Dahlia*, he traces these first, faltering steps, which were followed by a veritable explosion of interest and enthusiasm. Writing in 1838, he claimed that, 'No plant, or tribe of plants, of the most acknowledged beauty, or the most extensive variety of form and colour, has ever excited so much interest and attention, or been so successfully and universally cultivated by British florists or horticulturists.' He was astonished at the speed with which dahlias spread from being a rarity for the rich to a flower for the masses. 'Eight or ten years since,' he wrote, 'it was considered a perfectly novel sight, to witness Dahlias with double flowers in the garden of a tradesman or cottager; but…it is now quite as rarely that we see or meet with a cottager's garden, which does not contain at least a few good Dahlias; and many possess plants of first-rate sorts.'

The entire country had gone dahlia mad, and there were dedicated dahlia shows from Sheffield to the Isle of Wight. By the end of the Victorian era, it's said, there were around ten thousand varieties to choose from, but astonishingly, of all those only three survive today: the breast-beating 'Kaiser Wilhelm' and 'Union Jack', and 'White Aster', a handsome pompon type. So where did the other 9,997 varieties go?

They were, it seems, mainly a victim of their own success. Despite their dour reputation, Victorians adored brilliant, blowsy flowers, planting entire dahlia walks whose clashing colours must have been quite something to see. But even in their pomp, not everyone was a fan. Slowly floral fashions turned against them, towards more naturalistic planting with a softer, more restrained colour palette, of the kind captured so well in the wispy watercolours of Helen Allingham. As dahlias' popularity declined, there was no reason for commercial growers to maintain their stocks, and one by one the old varieties died out.

It wasn't until after the First World War that dahlias started to come back into fashion, at least partly thanks to the work of the British botanist and proselytizer William John Cooper Lawrence of the John Innes Horticultural Institution at Merton Park in Surrey. During the 1920s he created hundreds of experimental dahlia hybrids, one of which proved to be hardier and easier to grow than older varieties, and it is from Lawrence's hybrid that most modern cultivars descend.

To finish our disquisition on dahlias, there's a wonderfully withering dahlia-related put-down in a (presumably apocryphal) anecdote about the twentieth-century novelist Angus Wilson, who, the story relates, was a dreadful snob. Wilson had managed to ingratiate himself with a horticulturally minded duke, who was proudly showing Wilson around his ancestral acres. Spotting the ducal dahlias, the ever-waspish Wilson declaimed, 'I always think dahlias are so *suburban*, don't you, your Grace?' To which the duke replied, 'Sadly, Mr Wilson, I am not acquainted with the suburbs.'

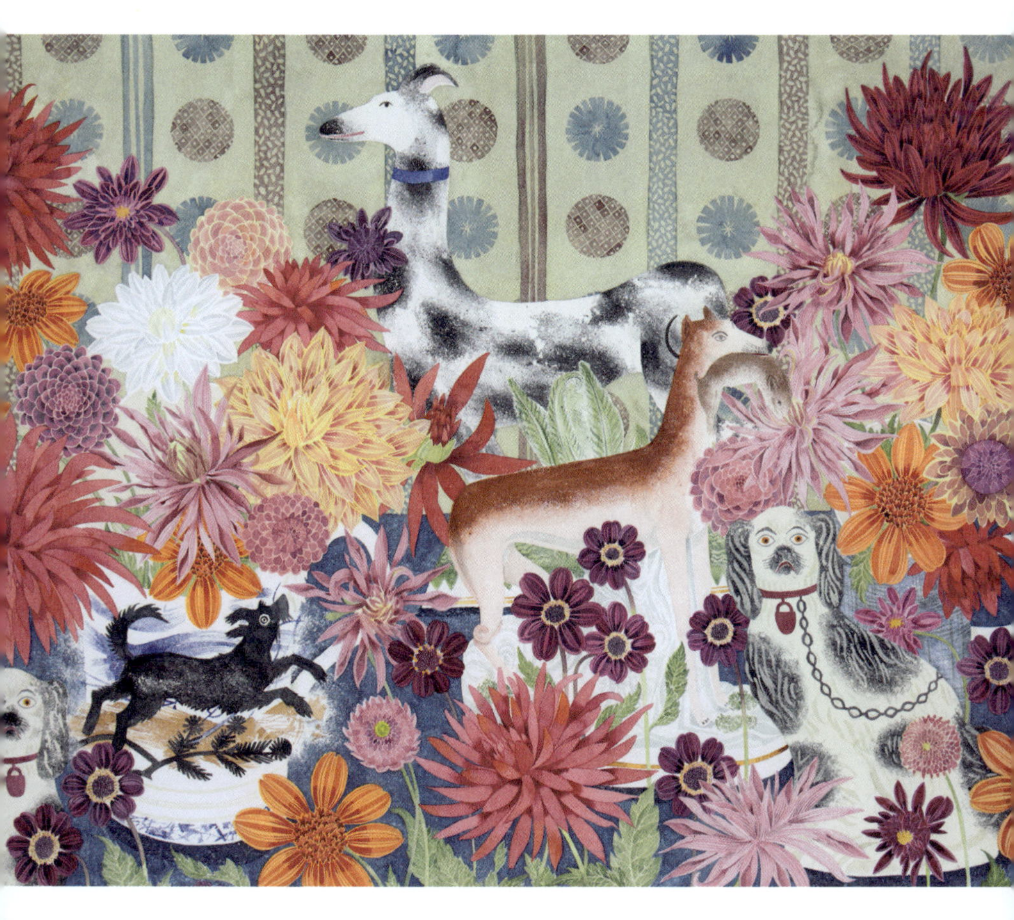

Dahlias and Dogs
Watercolour, 2022

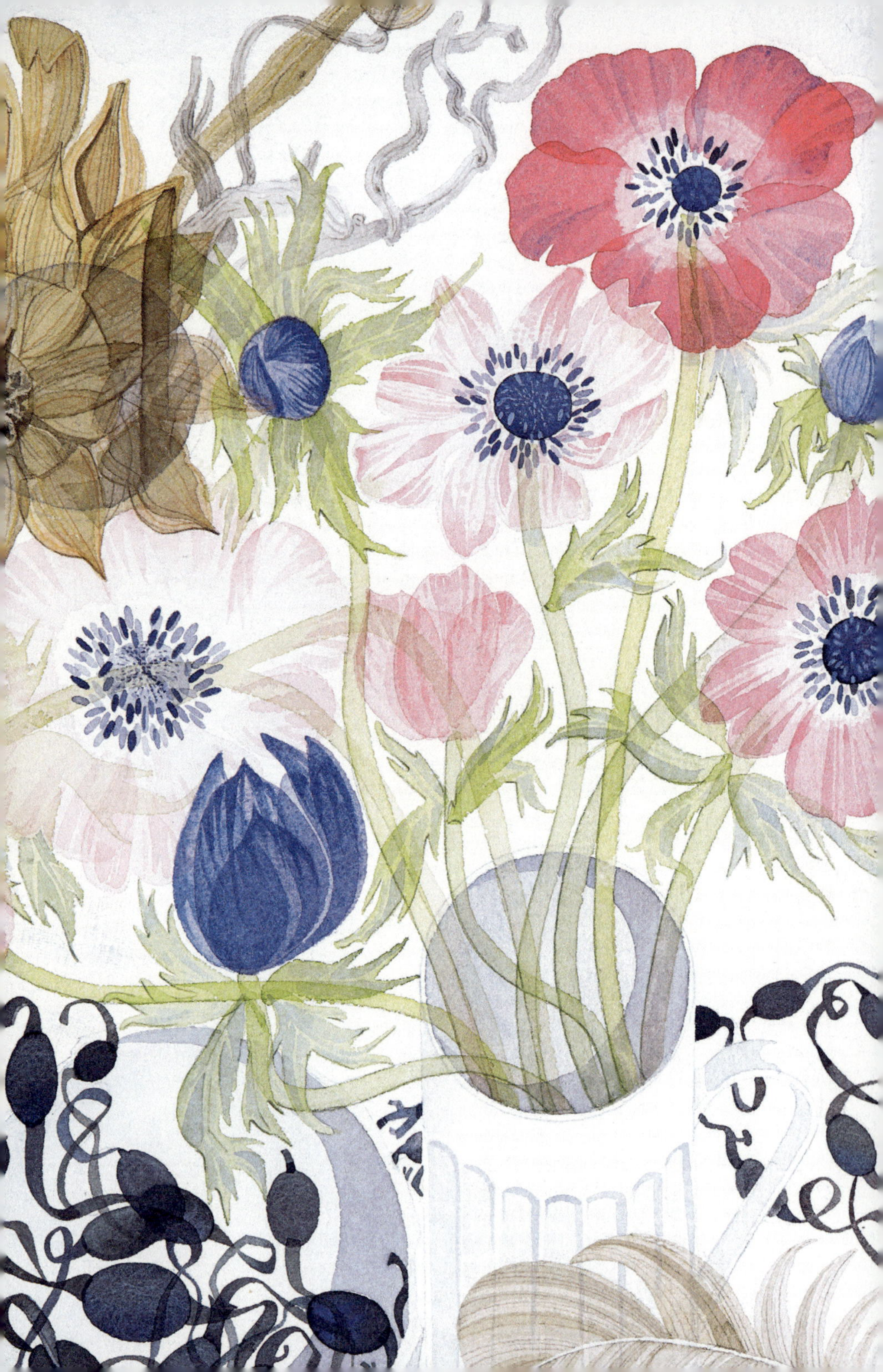

6

ANEMONES

The first time I saw *Anemone coronaria* growing in the wild they stopped me in my tracks. It was high in the Taygetos mountains of southern Greece, and although it was late in their season and some of them were looking a tad tatty, their colour was astonishing: the most eye-popping scarlet I'd ever seen, with an inky-black central boss of stamens, and ruffs of bright green leaves. Their colour is so intense that cameras struggle to capture it, so they really have to be seen to be believed.

Poppy anemones or windflowers, as they used to be known, encircle the entire Mediterranean and reach as far east as Iraq, growing wild often in the light shade of olive groves, and can withstand pretty arid conditions. They come in a range of colours, including lilac, pink, white and blue, although their scarlet forms are the most instantly transfixing, especially when they have an inner ring of

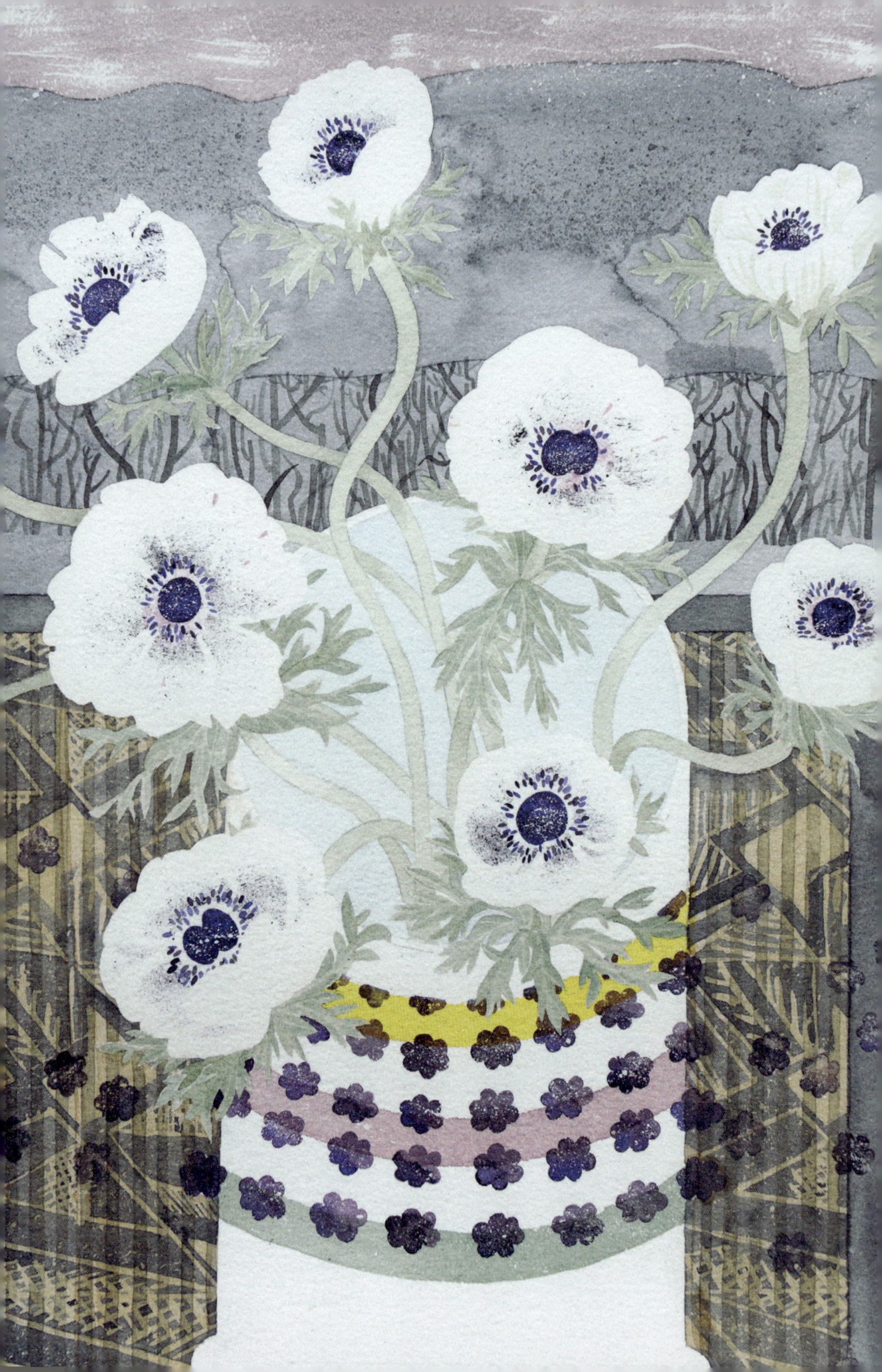

white around the central boss. At first glance they can be confused with another wild anemone, *A. pavonia*, but the latter has narrower and more numerous petals, and its native range is restricted to the north-eastern Mediterranean. Oddly, although they sometimes grow together, the two species decline to interbreed.

A. coronaria has an intriguing relationship with a family of scarabid beetles called the Glaphyridae, whose members appear to be its primary pollinators in the eastern Mediterranean, as well as a group of other large, bowl-shaped red flowers that botanists refer to as the 'poppy guild'; these include the common poppy, *Papaver rhoeas*, *Ranunculus asiaticus* (often sold as a cut flower) and *Tulipa agenensis*. The beetles fly from flower to flower in search of pollen, but also to find a mate, since female beetles seem to prefer the largest flowers with the most abundant pollen, and the males seem to be able to distinguish the largest flowers – which are most likely to be occupied by a female beetle – too.

A. coronaria grows from little wrinkled brown underground tubers, which are easy to pop in a pocket, and doubtless many early travellers, enraptured by the spectacular colour of the flowers, did just that, spreading it far and wide. It seems more than likely that it was cultivated in ancient Rome, and, according to the garden historian Maggie Campbell-Culver, it may well have been brought back to Britain as early as the twelfth century by crusaders returning from the Holy Land, where it grows in profusion. *A. coronaria* is the national flower of Israel, as well as one of the national flowers of Palestine. Its vivid red was associated with

Page 50: *Anemones and Artichokes* (detail)
Watercolour, 2018

the blood drops of Christ on the Cross, and Campbell-Culver believes the Knights Templar grew the plant at their preceptory at Ribston, near Harrogate – although one wonders how long a tender Mediterranean native would survive in North Yorkshire.

We can be sure, however, that it was blooming in English gardens by the mid-sixteenth century, since John Gerard included it in his famous *Herball* of 1597, describing a 'Great Double Windflower of Bithynia' (present-day northern Anatolia, Turkey). As the wide range of natural colours suggests, *A. coronaria* hybridizes readily, and gardeners took advantage of this to conjure up hundreds of cultivars – most of them more or less ephemeral. The earliest enthusiasts appear to have been aristocrats, diplomats and other wealthy collectors, who swapped plants with one another, often over long distances, and delighted in displaying their rare acquisitions in special narrow flowerbeds.

From the 1620s on, collectors and hybridizers became known as 'florists' (a term originally used to describe flower fanciers rather than flower arrangers), who held their own dinners and initially confined themselves to just seven plants: anemones, auriculas, carnations, hyacinths, polyanthus, ranunculus and tulips. Just like gardeners today, they loved anything out of the ordinary, such as double flowers (like John Gerard's 'Great Double Windflower of Bithnyia') or flowers with striped or spotted petals. Charles I's official botanist, John Parkinson, was a big anemone fan. 'So pleasantsome and delightsome flowers that the sight of them doth enforce an earnest longing desire…to be the possessor of some of

Pink Anemones (detail)
Watercolour, 2018

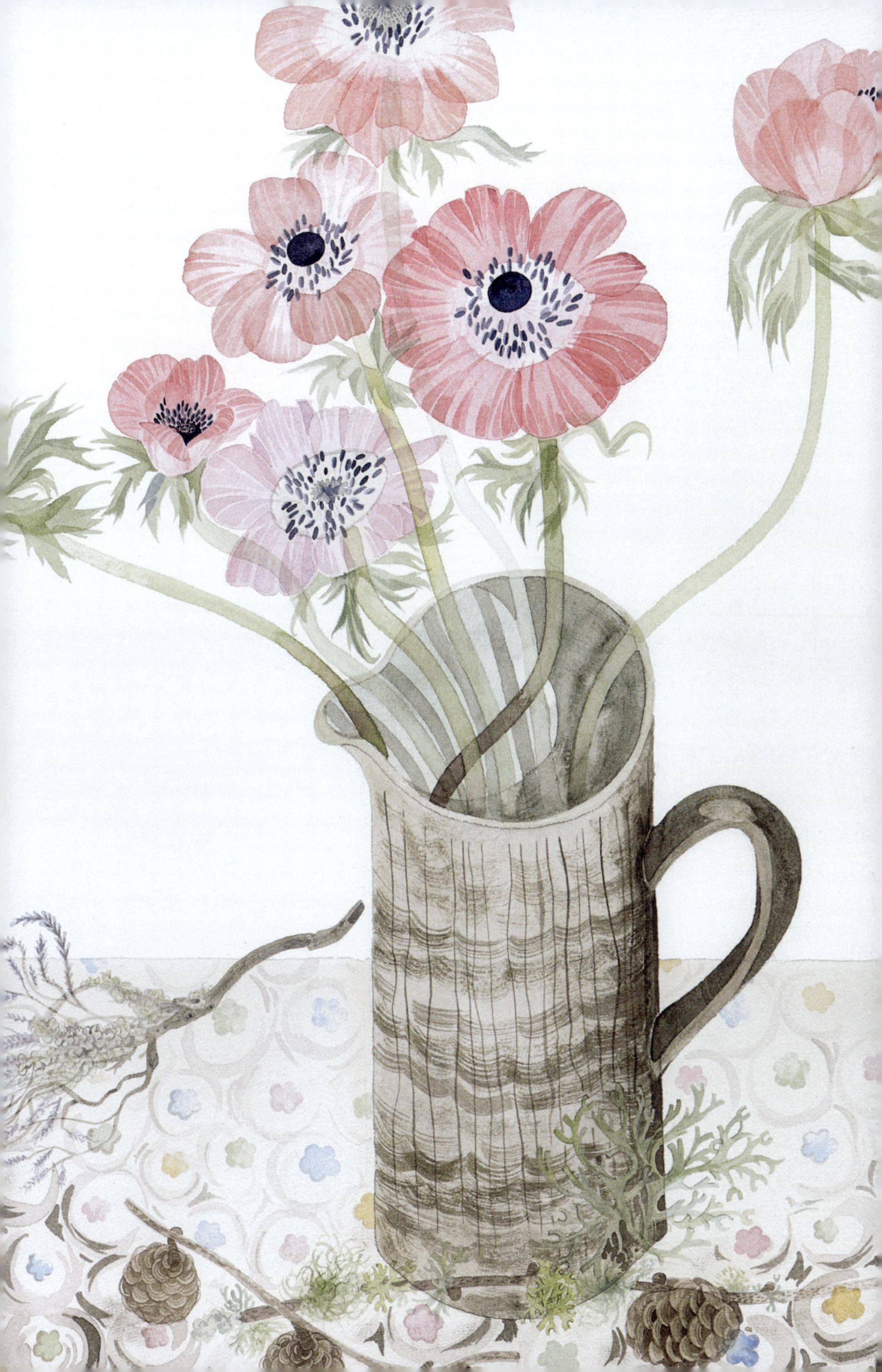

them,' he wrote in his treatise on gardening *Paradisi in Sole Paradisus Terrestris: A Garden of All Sorts of Pleasant Flowers Which Our English Ayre Will Permit to Be Noursed Up* (1629), and he grew them in his own London garden in Long Acre, Covent Garden.

Although anemones seem to have fallen out of fashion in Britain during the eighteenth and nineteenth centuries, they continued to be popular in the Netherlands, France and Ireland. Many of the large single-flowered *Anemone coronaria* cultivars sold today belong to the so-called De Caen group, which gets its name from the French town that, in the mid-nineteenth century, became famous for the anemones grown there by one Madame Quétel, a well-known widow whose display of cut anemones at the Cherbourg flower show in 1848, and at later shows, caused a sensation. Although her cultivars apparently included a mix of single, semi-double and double flowers, by the end of the First World War 'De Caen' had narrowed to refer only to single-flowered forms.

The well-known 'St Brigid' double forms of *A. coronaria* originated in the 1880s thanks to the work of a talented Irish gardener called Alice Lawrenson (1841–1900), who lived in Howth at the northern end of Dublin Bay. She named her double-flowered seedlings after St Brigid or Brigid of Kildare, the patroness saint of Ireland, and went on to introduce new varieties of Christmas roses and daffodils (including *Narcissus* 'Lucifer'). Her anemones became so popular that they gave their name to any double-flowered cultivars, and 'St Brigid' anemones continue to be grown widely today.

Among the single-flowered varieties, several of the most highly regarded examples originated in Holland, including 'The Bride', a lovely pure white form. It was introduced by the Haarlem-based company E. H. Krelage & Zn in 1870 and originally named *Anemone alba superbissima*, but by 1883 it was being marketed as 'The Bride'. Three new cultivars appeared in the late 1920s: the scarlet 'Hollandia', the magenta 'Sylphide' and the vividly violet-blue 'Mr Fokker'. The first was introduced in 1927 by Vrugt of Hillegom, south of Haarlem, a firm that specialized in anemones and ranunculus, while the other two are first mentioned in the autumn catalogue of C. G. van Tubergen of Haarlem from 1928. I'm guessing that 'Sylphide' was named after the ballet by Delibes, but which famous Fokker inspired the last variety remains a mystery; could it be the Dutch founder of the aircraft company, Anthony Fokker? We may never know.

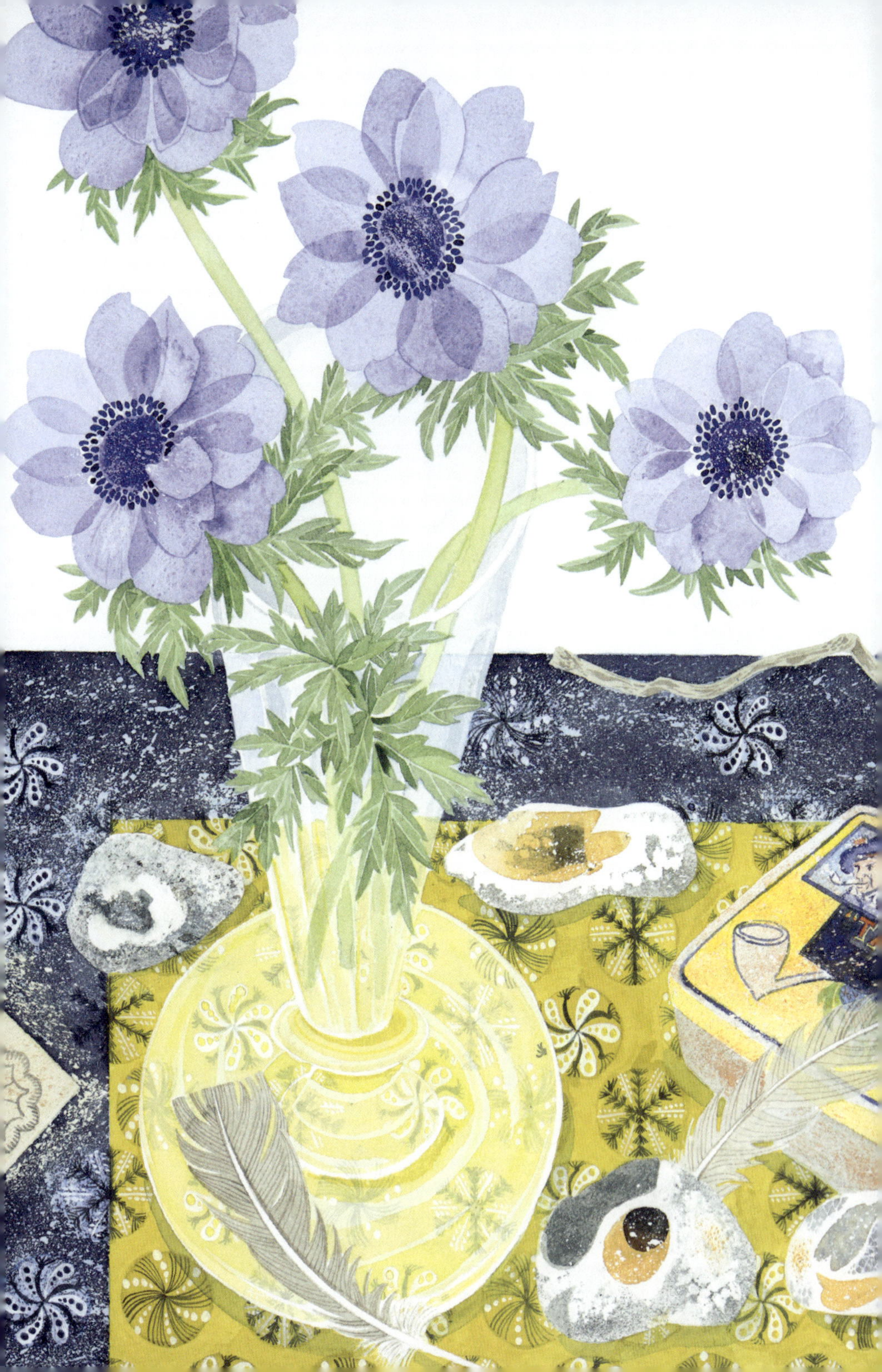

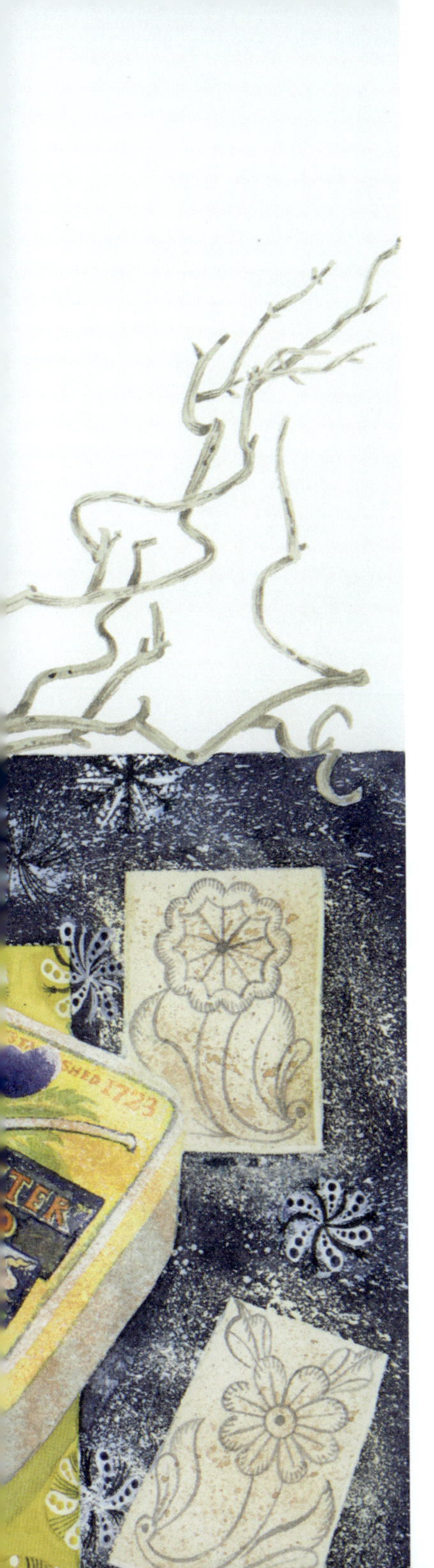

Anemones and Tam O'Shanter Tin
Watercolour, 2022

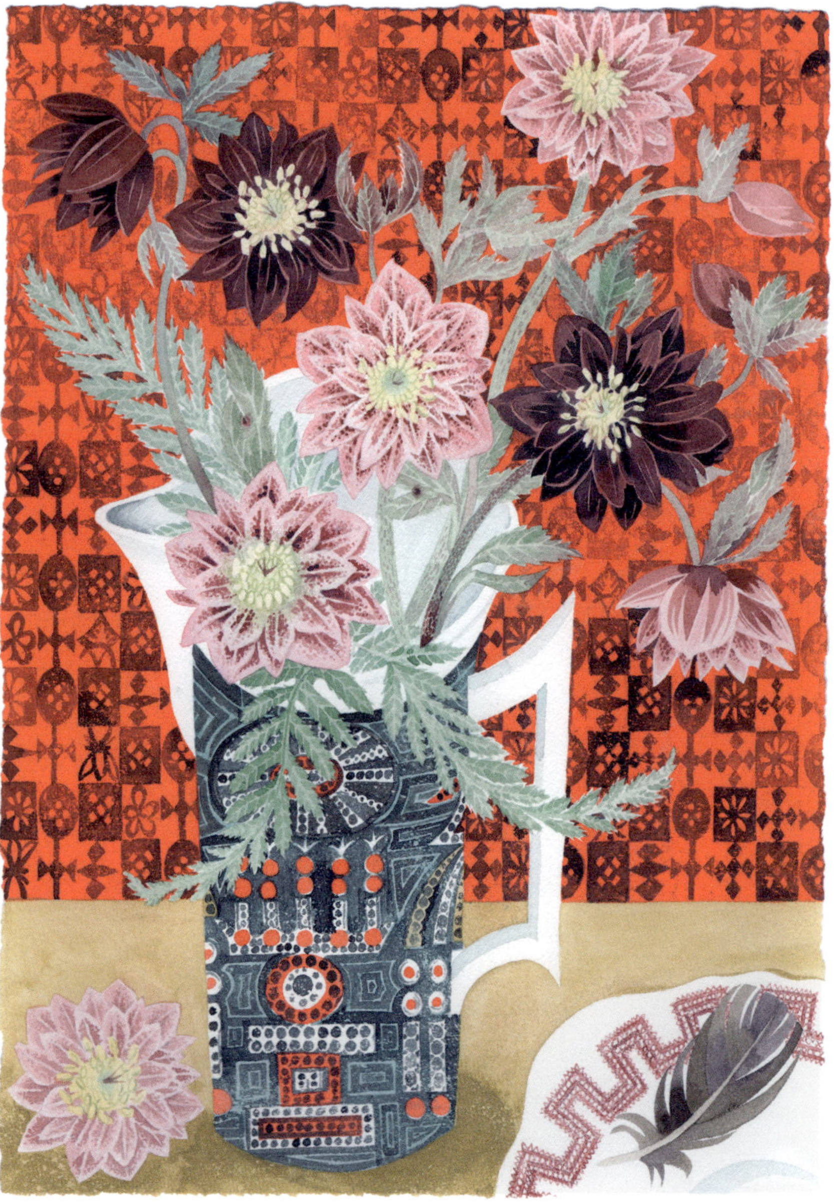

7

HELLEBORES

When I was growing up in our front garden we had a large Christmas rose, which occasionally flowered when it was supposed to. Even as a child I was puzzled by its name, since it didn't look much like the other roses that we had, and what was it doing coming into flower in December and January anyway? Of course, I now know that Christmas roses are a species of hellebore, and nothing to do with actual roses; in fact, they're in the same family as buttercups and anemones. There are around fifteen species of hellebore, most of which are native to Europe, from Britain in the west to Turkey in the east, plus one species that's found only in China.

Hellebores and 'Magic City' Jug
Watercolour, 2024

They're tough, handsome plants, most of which flower in the winter and are tolerant of shade and drought, making them extremely useful in the garden – not only for flowering when hardly anything else is out, but also because some of them will grow in dry shade, where hardly anything else will thrive. Our own native species, *Helleborus foetidus*, is perfectly happy under trees and hedges, and its tall, multi-headed pale green flowers and dark, shiny leaves make a strong impression at any time of year. By contrast the Christmas rose, *H. niger*, needs full sun and a sheltered spot if it is to do well.

Most garden hellebores, however, are hybrids, often between *H. orientalis* (which originates in Greece and Turkey) and other species or other hybrids. Today there are hundreds, if not thousands of varieties to choose from, but it wasn't always that way. Until the 1960s, most hybrid flowers came in an uninspiring range of muddy pinks and greens, but a series of hellebore heroes changed all that. The pioneer, Helen Ballard, had a fascinating life. Born in 1908, she married Peter Wilson, youngest son of Sir Mathew 'Scatters' Wilson, who worked for MI6 during the Second World War before joining Sotheby's and rising to become one of its most successful chairmen. The couple had two children, but divorced in 1951 after it became apparent that he really preferred men.

Helen remarried that same year, to a farmer called Philip Ballard, whose father had established the Old Court Nursery at Colwall, in the shadow of the Malvern Hills in Worcestershire. Living at Mathon nearby and with time on her hands, Helen became fas-cinated by the hellebores introduced by the nursery's new owner,

Percy Picton, and began experimenting with some of the species and hybrids that he grew. From the 1960s to the 1980s she developed new hybrids of outstanding quality, often with striking colours, including improved Christmas roses and the first really dark hellebore, 'Ballard's Black'. Her hellebores became legendary (and expensive), and, although she died in 1995, many of today's finest cultivars can trace their ancestry back to her original hybrids. Old Court Nursery, incidentally, is still going strong, and Ballard's home at Old Country Farm in Mathon is now run as a bed and breakfast, with hellebores still very much in evidence in the garden.

Ballard's pioneering work was carried forward by Elizabeth Strangman at Washfield Nursery near Hawkhurst in Kent, which she ran between 1968 and 1999. Strangman raised many fine hybrids at Washfield, but also travelled widely and collected seed from native species in the wild, including two double-flowered forms of *H. torquatus* that she found in Montenegro, from which she raised some of the first modern double-flowered hybrids. She developed unusually strong, clear colour forms, including apricot and yellow, as well as the now popular 'picotee' cultivars, which have ruffled petals; and she shared her knowledge in *The Gardener's Guide to Growing Hellebores* (1993), which is a classic of its kind.

Strangman's and Ballard's hybrids were further developed and commercialized by Robin White at Blackthorn Nursery in Hampshire and John Massey at Ashwood Nurseries in the West Midlands, the latter of which continues to offer superb varieties today. With so many cultivars now on offer, it's worth taking the trouble to visit a specialist nursery rather than a generalist garden centre, since the former's plants are more likely to do well in the open garden, and the staff will be able to give you really informed advice on how to ensure the plants thrive. But in the end, the only real risk is falling head over heels in love with them and becoming (to use a word coined by the celebrated gardener and writer Christopher Lloyd) a helleBORE.

Hellebores and Snowdrops (detail)
Watercolour, 2024

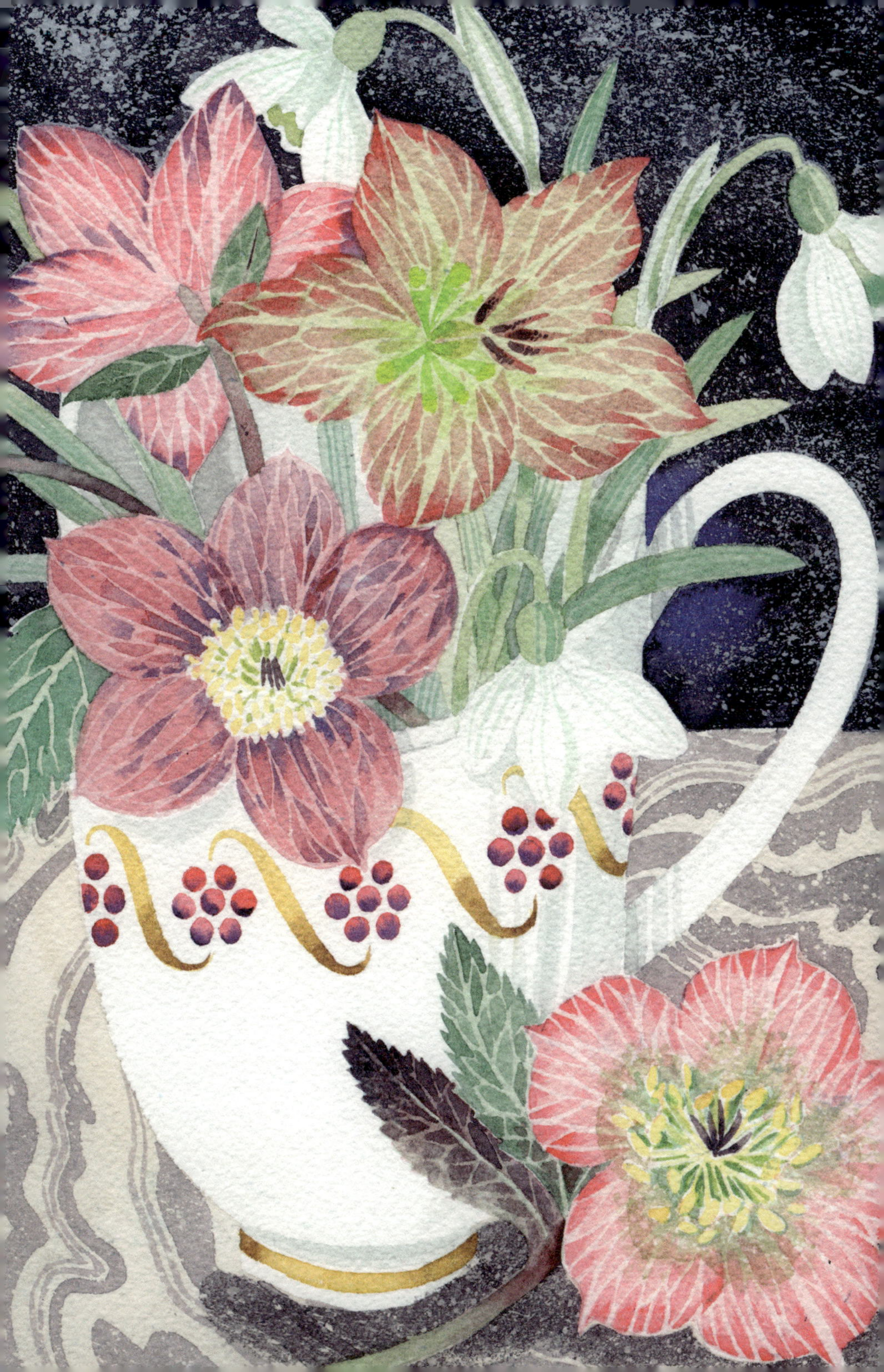

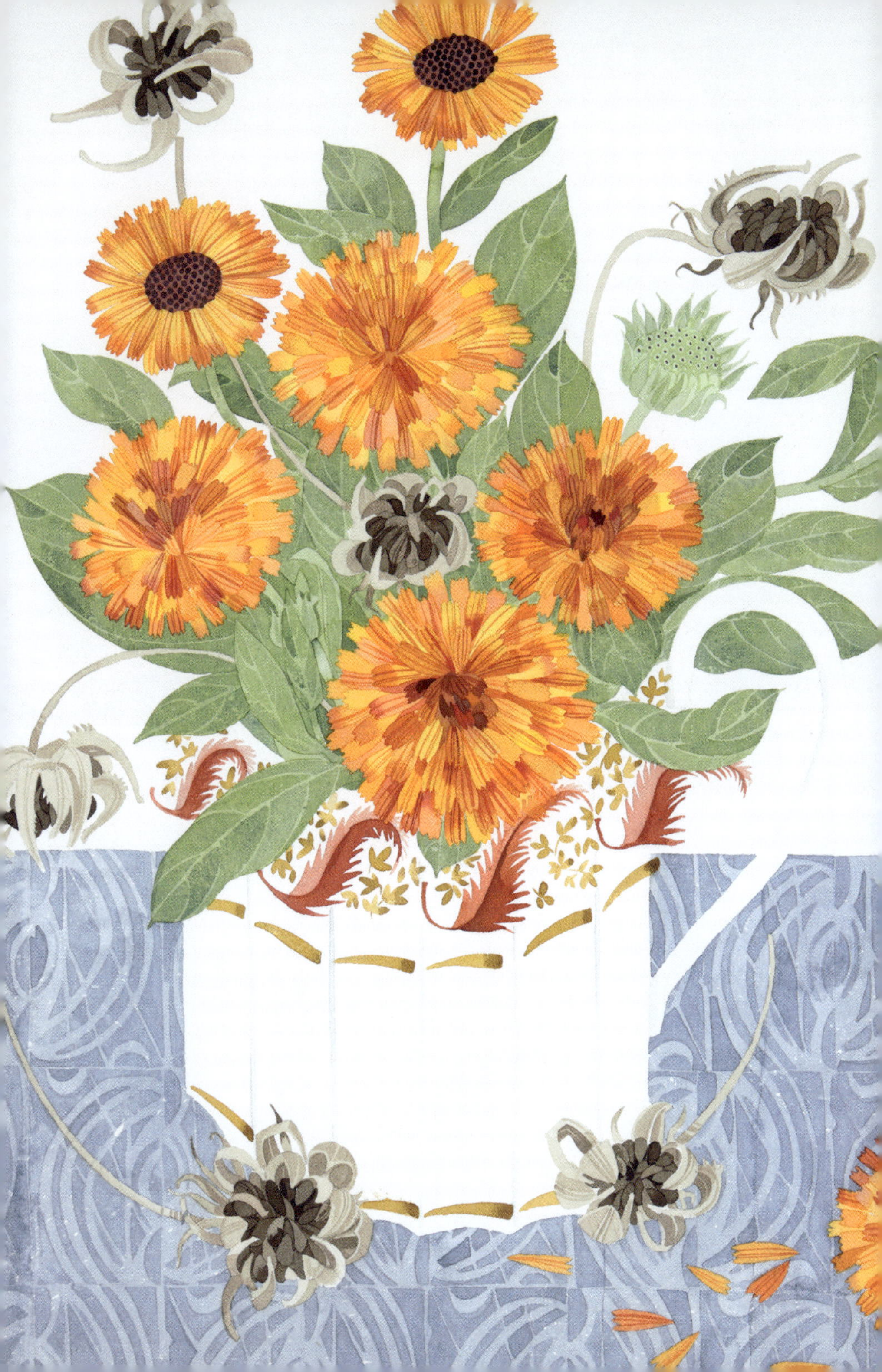

8

CALENDULAS

Using Latin names for plants might seem a bit pretentious, as well as mystifying, but marigolds demonstrate the perils of common names. We grow three kinds of marigold in our gardens – French marigolds, pot marigolds and African marigolds – but two of them are completely unrelated to the other, and (apart from their flowers being mostly orange) look pretty different, too. French and African marigolds have crinkly flowers, finely divided leaves and a fetid smell, while pot marigolds have simple daisy-like flowers and sticky leaves. To add to the confusion, neither French nor African marigolds come from France or Africa, but rather from Mexico, whereas pot marigolds seem to have originated around the Mediterranean (possibly centred on Spain). In botanical Latin, both 'French' and 'African' marigolds belong to the genus *Tagetes*, while pot marigolds are in the genus *Calendula*; no confusion there.

Calendula officinalis (to stick with the Latin name for simplicity) is one of those intriguing plants that have been cultivated for so long that it's possible its original ancestor is extinct – as seems to be the case with onions and broad beans – and that the plant we know is an ancient hybrid. It's quite similar to the field marigold, *C. arvensis*, which grows right around the Mediterranean, but it generally has larger, more brilliantly orange flowers, and it's this vibrant colour that makes it such an appealing garden staple. The plant historian Maggie Campbell-Culver believes *C. officinalis* has been grown in Britain since the twelfth century, and it is one of the easiest plants to grow, except in the coldest conditions; in common with most annuals it comes up quickly from seed, and can start flowering within weeks, as long as the weather is warm enough. With only a year to complete its life cycle, it also sets a lot of seed, so once you've got calendulas they perform more like an unusually attractive weed, popping up in any sunny corner (luckily, they're also easy to pull up).

This marigold's Latin name, like so many others, was assigned to it in the eighteenth century by the great Swedish botanist Carl Linnaeus. He chose *Calendula* as the genus name in recognition of the fact that, in warm climates, pot marigolds could feasibly be in flower every month of the year – *Calendula* comes from the same Latin root, *kalandae* (meaning the first day of the month), as the word calendar. Linnaeus applied its species name, *officinalis*, to many other plants that had been widely used in herbal medicine and cookery; other examples include *Salvia officinalis* (sage), *Melissa officinalis* (lemon balm), *Pulmonaria officinalis* (lungwort) and *Borago officinalis* (borage).

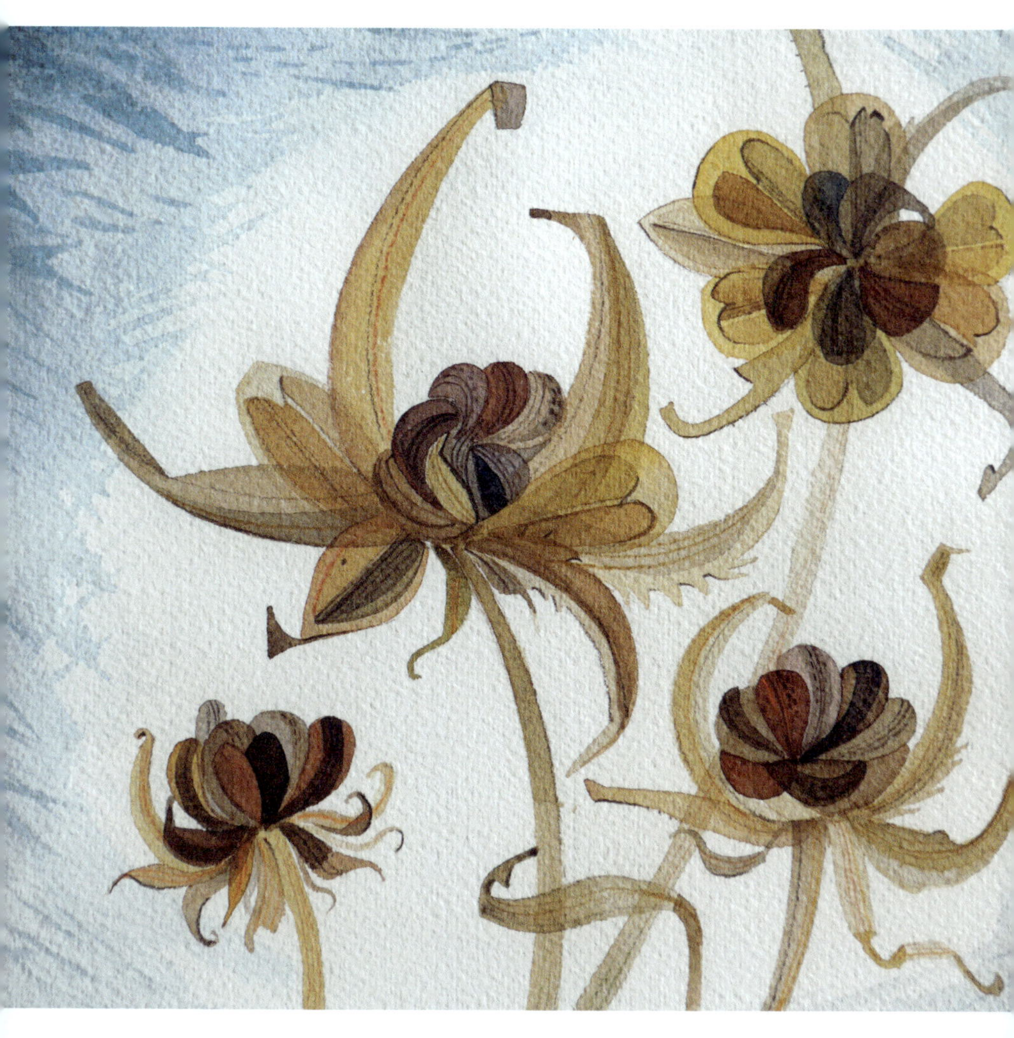

Page 66: *November, Last of the Marigolds* (detail)
Watercolour, 2022

Marigold's common name, ironically, is more complex. The 'gold' bit is easy enough; there aren't many flowers so shiny and golden in colour. The 'mari' bit is generally assumed to refer to the Virgin Mary, but according to the scientist Rexford Talbert of the Herb Society of America, this explanation 'is relatively recent in English history and is but another case of religious opportunism. As my professors used to say, an answer that is both succinct, tidy and untrue.' Talbert offers an alternative etymology, which is that 'marigold' is a corruption of the Anglo-Saxon word *meargealla*, which was originally applied to the native yellow marsh marigold or kingcup, *Caltha palustris*. His theory is that when *Calendula officinalis* was introduced, perhaps as early as the twelfth century, people extended the name *meargealla* to include calendulas too. After *meargealla* had turned, over the centuries, into 'marigold', someone recognized the similarity between 'mari' and 'Mary' and started explaining its origins in religious terms. It's an engaging theory, not least because the truth is so often much messier and less obvious than long-cherished fabrications, but in the absence of solid documentary evidence either way, perhaps it's best to err on the side of agnosticism.

As with other popular garden plants, various marigold cultivars have been grown over the centuries, although they're not as varied or as numerous as, say, dahlias or daffodils. Perhaps the best is *C. officinalis* 'Indian Prince', which has tall, richly coloured flowers in dark orange to russet shades, with a dark brown centre. 'Snow Princess', by contrast, is a creamy white and half the height of 'Indian Prince'. For those who find classic calendula flowers too glaring, there are softer-coloured variants, such as 'Art Shades' and 'Playtime Mix', all of which are equally easy to grow.

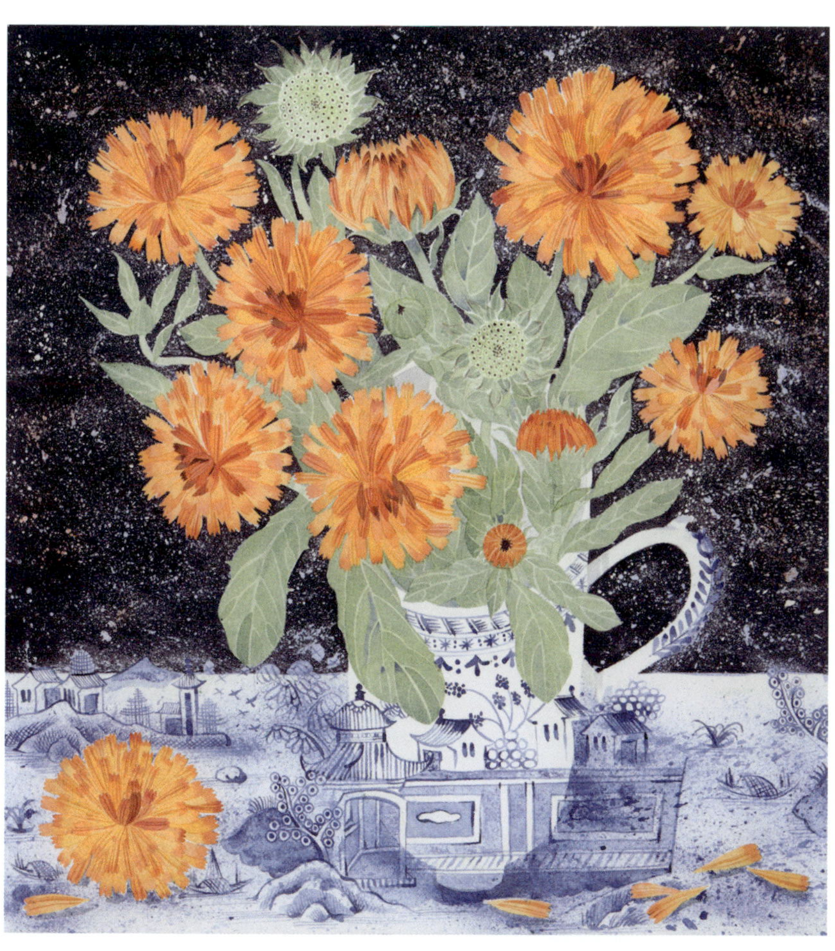

Marigolds and China Cup
Watercolour, 2022

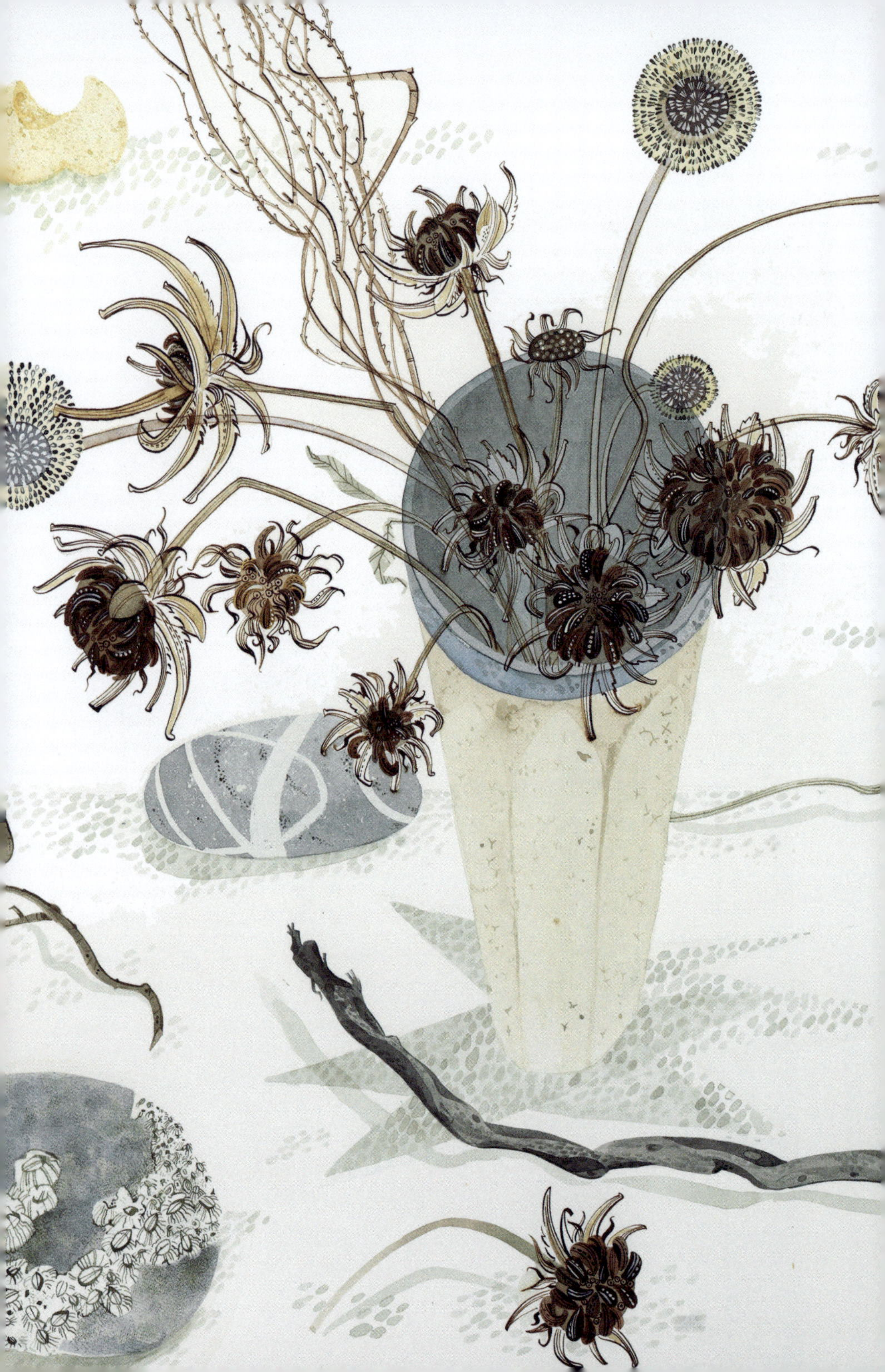

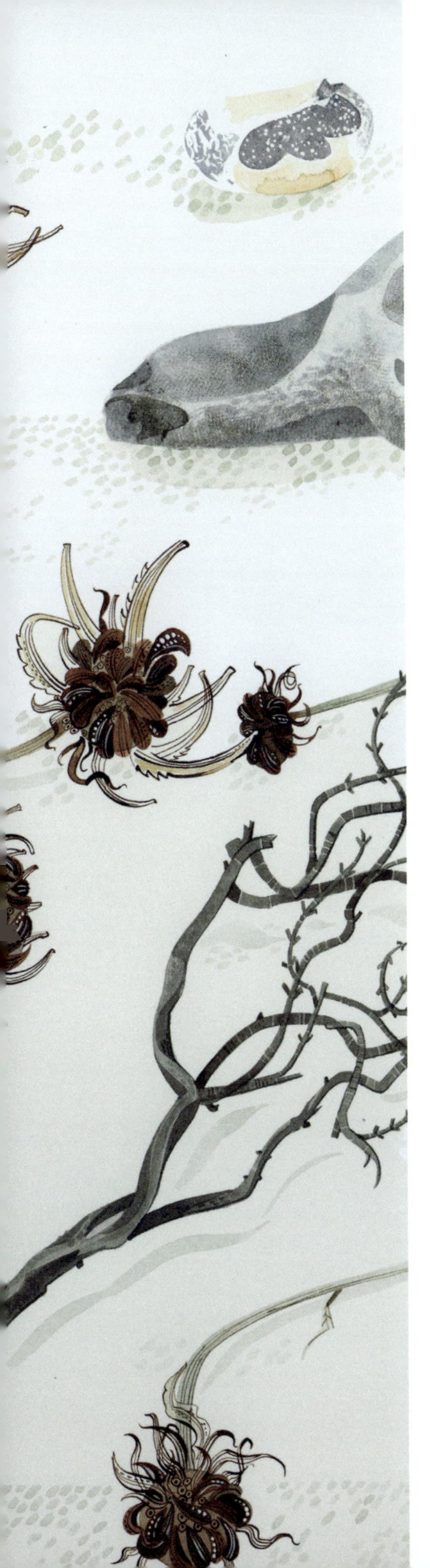

Twisted Stem, Calendula and Pebbles
Watercolour, pen and ink, 2019

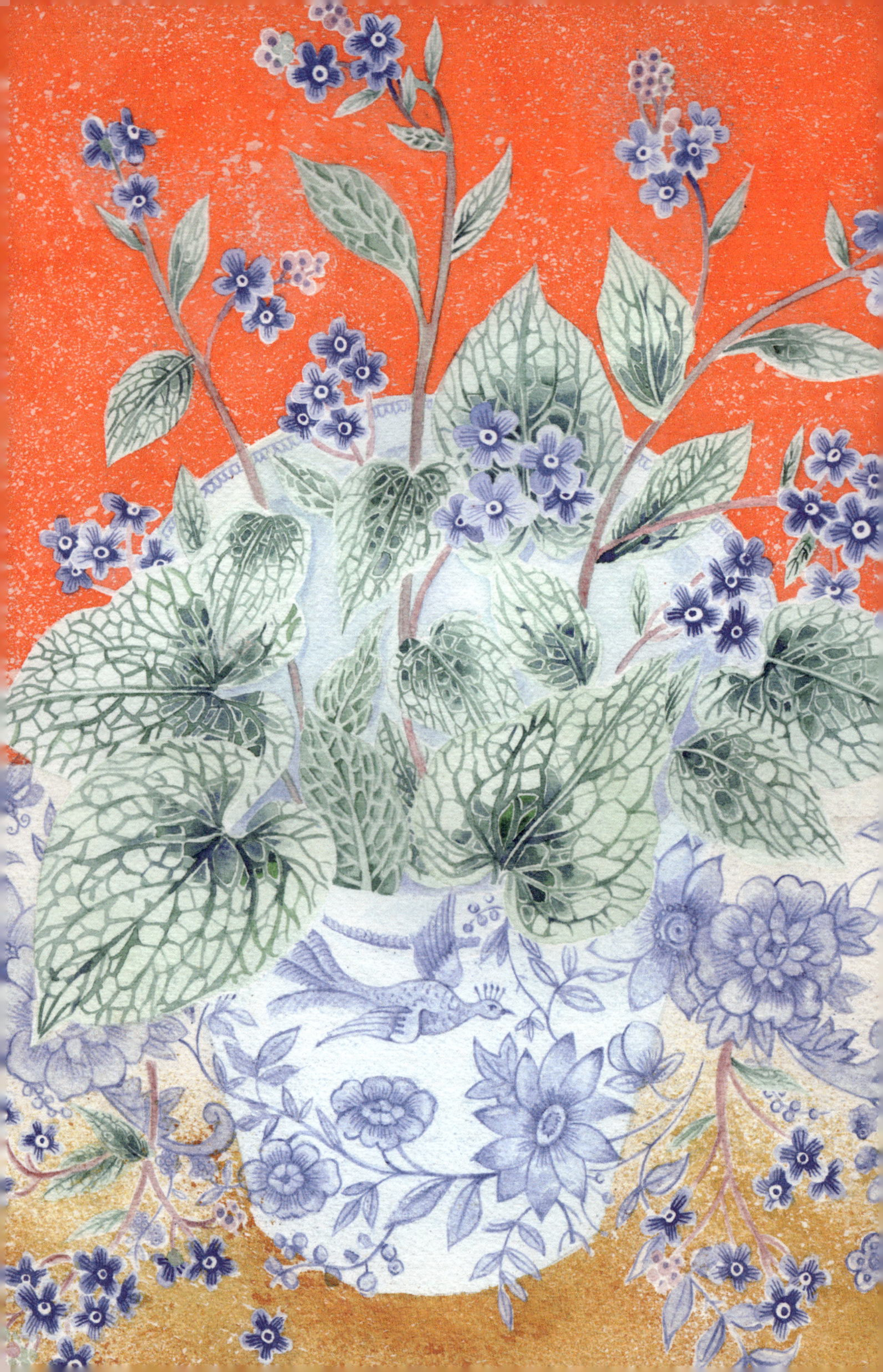

9
BRUNNERA

The profusely blooming forget-me-not (*Myosotis sylvatica*) is a staple of old-fashioned cottage gardens, with its sprays of sky-blue flowers and a habit of self-seeding all over the place. These are pretty plants, but they have their drawbacks. They're short-lived, for starters, and their small, rather hairy leaves are susceptible to mildew in warm weather. So it's easy to see why *Brunnera macrophylla*, the so-called Siberian bugloss, has become so popular since it was introduced into British gardens in 1713. Its flowers look very like forget-me-nots (and indeed it belongs to the same family, the Boraginaceae, along with borage and comfrey), but it has larger, more distinctive leaves – heart-shaped, with pointed tips – and grows as happily in semi-shade as in sun, although it doesn't like drying out. Unlike forget-me-nots, it's a long-lived perennial, and although brunnera

Brunnera in Pheasant Cup
Watercolour, 2024

rarely sets seed, it spreads by means of long, swollen roots just below the surface of the soil, and makes a useful ground-cover plant.

As is so often the case, the common name of Siberian bugloss is misleading, since *B. macrophylla* actually originates in the Caucasus mountains and Turkey. Admittedly, one of the two other brunnera species, the aptly named *B. sibirica*, does come from Siberia – but that is rarely grown in gardens. Over the years several cultivars of *B. macrophylla* have been introduced, including the white-flowered 'Betty Bowring' and numerous types with variegated leaves. One of the most alluring, 'Jack Frost', which has silvery leaves veined with and edged in green, was introduced in the year 2000 by Walters Gardens Inc. of Zeeland, Michigan, where it had been discovered among the progeny of a spotty-leaved cultivar called 'Langtrees'. More novelties are bound to come, although I still think the original plain-green species is hard to beat.

Striped Cups with Spring Flowers (detail)
Watercolour, 2016

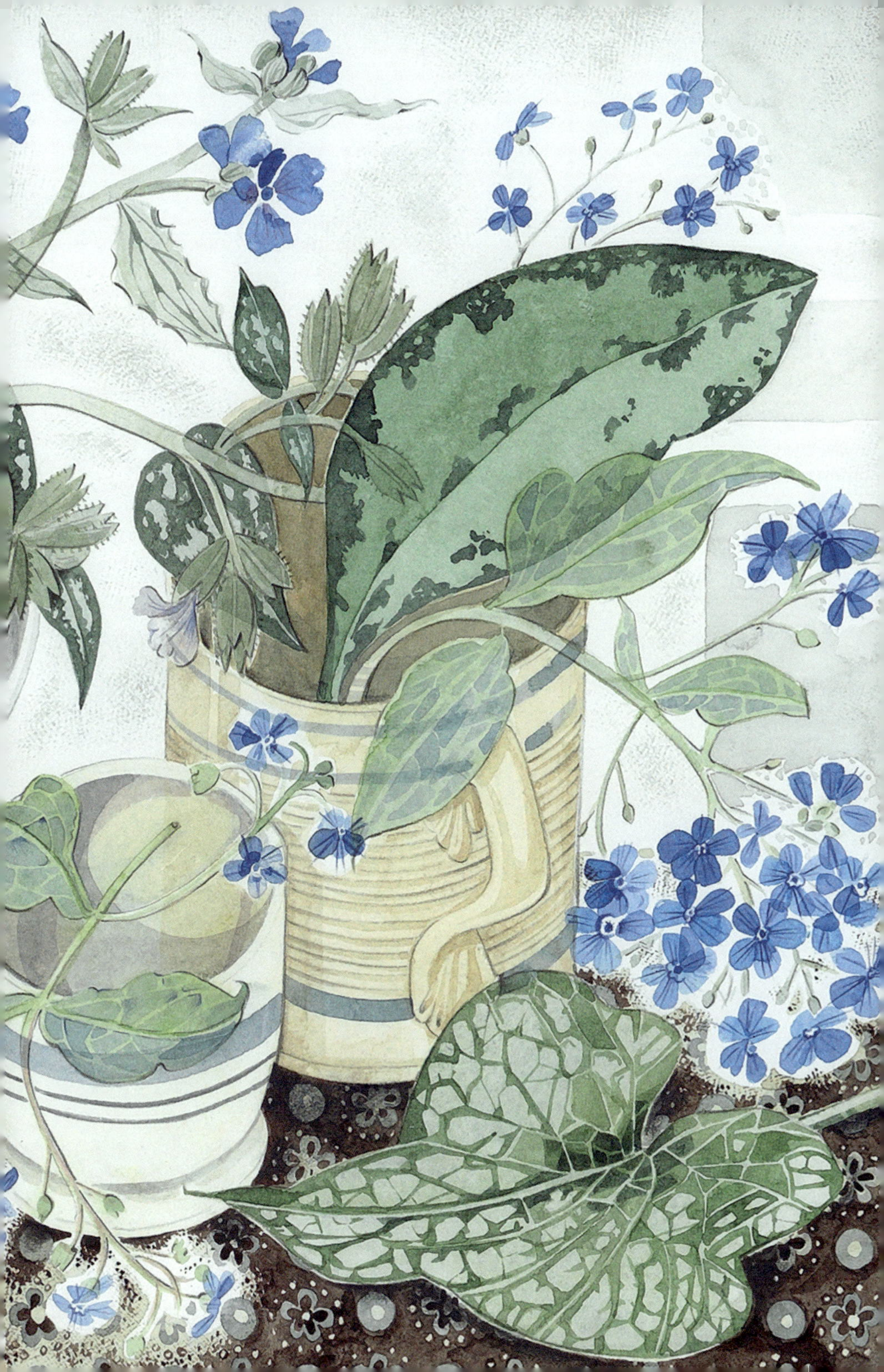

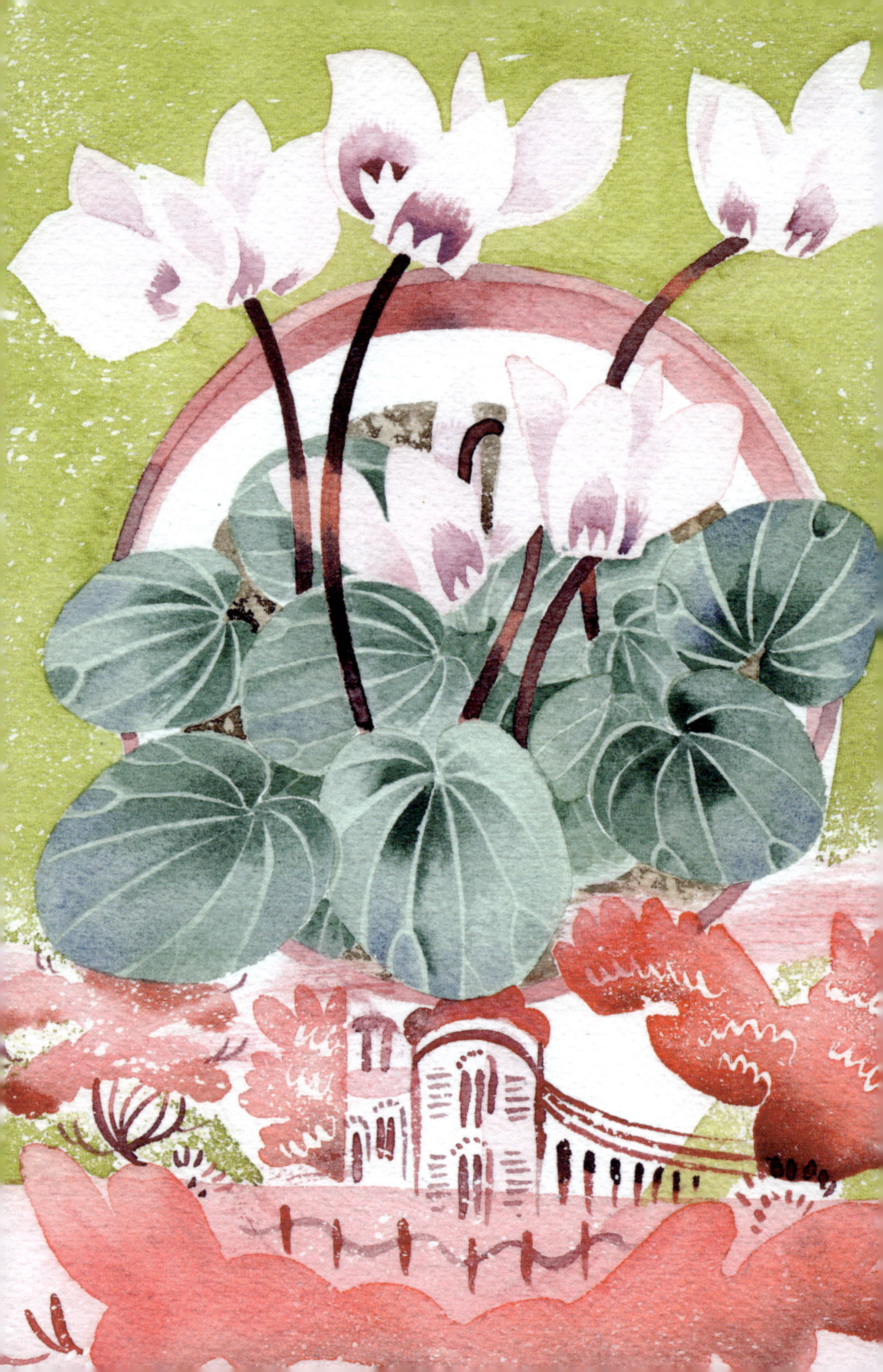

10

CYCLAMEN

With their diminutive flowers and exquisitely patterned leaves, cyclamen look like delicate, difficult plants to grow – but they're actually surprisingly hardy, as long as they don't get too wet. They come from a family of about twenty-four species, most of which grow around the Mediterranean, and – as have many other Mediterranean plants – they have adapted, over millions of years, to a scarcity of water. In common with carrots and potatoes, cyclamen rely on a bulb-like structure called a tuber (in effect a swollen stem), which stores enough energy in the form of starch to see the plant through the longest summer droughts. These roughly circular, surprisingly hard brown tubers can grow to the size of flattened footballs in the wild, and probably gave cyclamen their name: *kyklos* was the ancient Greek for circle, which became *kyklaminos*, then *cyclaminos* in Latin.

The first mention of cyclamen being grown in Britain comes from the late sixteenth century, and by 1648 two species were in residence at the Oxford Botanic Garden – probably what we now know as *Cyclamen coum* and *C. hederifolium*, which are still the most popular garden cyclamen today.

Winter-flowering *C. coum* originates in the mountainous regions along the eastern edge of the Mediterranean, from Turkey as far south as Israel, but also around all but the most northerly coasts of the Black Sea. In the wild it's happiest in forest and woodland, where it's shaded from strong sunlight and has rich leaf mould to grow in, and unlike many species it enjoys relatively damp summers and wet winters. That, of course, suits it admirably to the British climate, helped by the fact that it's frost-hardy to -10°C (as long as it's not in waterlogged ground); no wonder it has been popular for so long.

As with other cyclamen, disparate populations of *C. coum* often have wildly varied leaves, which appear in autumn and die down again when the weather gets warm the following year. They can be heart-shaped or almost round, and come in a wide array of colours and patterns, from plain green to silver-grey, often with symmetrical patterning in darker or lighter greens. Yet another advantage is that their pretty flowers, which range from white through pale pink to deep magenta, bloom from December to March, when pretty much everything else is dormant or dead.

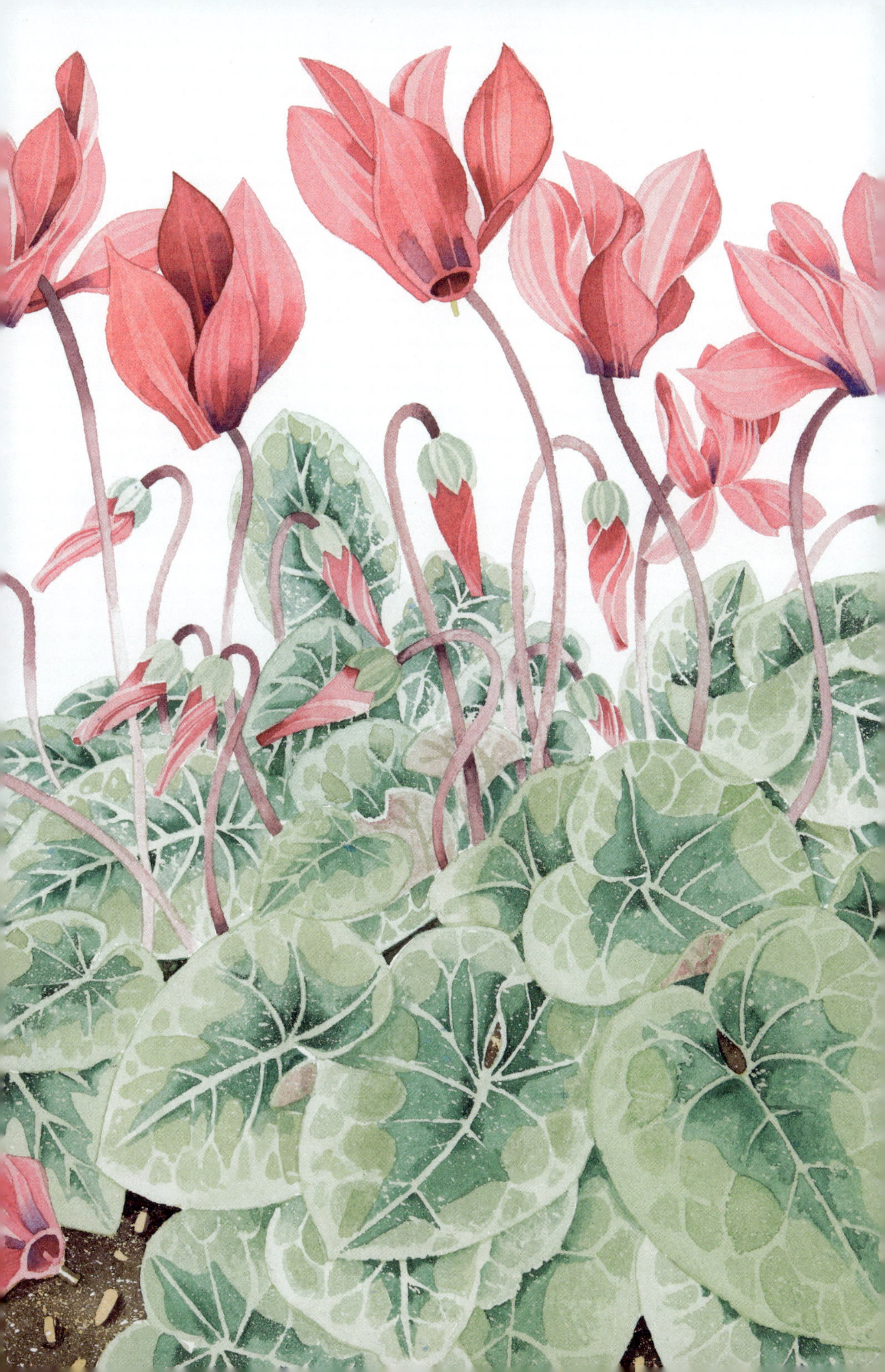

C. hederifolium is if anything even more adaptable, as its wide natural range suggests; it can be found growing in shaded spots from southern France and Switzerland through the whole of Italy and the Balkans to the west coast of Turkey. As its name implies (at least to those with a smattering of botanical Latin), *C. hederifolium* has ivy-like leaves – or at least the commonest type does, for its leaves are also remarkably variable, ranging from widely heart-shaped to long and pointed, and in colour from pale silver to almost begonia-like patterns of light and dark green.

Another woodland-lover, it is even hardier than *C. coum*, and once it gets established it naturalizes well, sets seed and can cover large areas of ground, producing rose-pink flowers in the autumn. One of the most wonderful sights I've ever seen was just off the coast of southern Greece, on a tiny island that was almost entirely covered with huge colonies of sugar-pink *C. hederifolium*, growing under pine trees and among the sharp rocks that fringe the sea.

Look closely and cyclamen have other attractions, too. Some are sweetly fragrant, although getting near enough to smell them can involve prostrating yourself on the ground (a pose that will be familiar to ardent botanists). Their flower buds are tightly furled in a spiral, like an umbrella with the tip pointing down, but when they open the petals twist upwards, back on themselves, making the flower look upside down. If you happen to be watching a flower bud at the right time and are very lucky, as I was one autumn morning just as the sun came out, you might even see it flip open, which takes only a second or two. It's a thrilling sight, as if the petals are on springs.

Later, if the flower has been successfully pollinated (by a wide range of small flies, bees and even moths), it transforms into a blackish, spherical pod filled with sticky-looking reddish seeds. Starting at the top, the flower stem then gradually curls over on itself, forming an ever-tighter spiral, which rolls the seed pod down until it sits on the ground. At this point the pod splits open, allowing ants, woodlice and other ground-dwelling invertebrates to rummage through the seeds, whose waxy coating they eat while – hopefully – dragging the seed to somewhere it can germinate.

Cyclamen are so naturally variable in leaf form and flower colour that most garden varieties are simply selected forms: plants that were originally collected in the wild with, say, silver leaves or white flowers, then propagated for sale by horticultural companies. There are at least twenty-five commercial cultivars of *C. hederifolium* (including 'No Name' and 'Daley Thompson') and a similar number of *C. coum*, so there's no shortage of choice for the garden.

Page 78: *Cyclamen in Lustreware Cup*
Watercolour, 2023

Page 81: *Cyclamen* (detail)
Watercolour, 2024

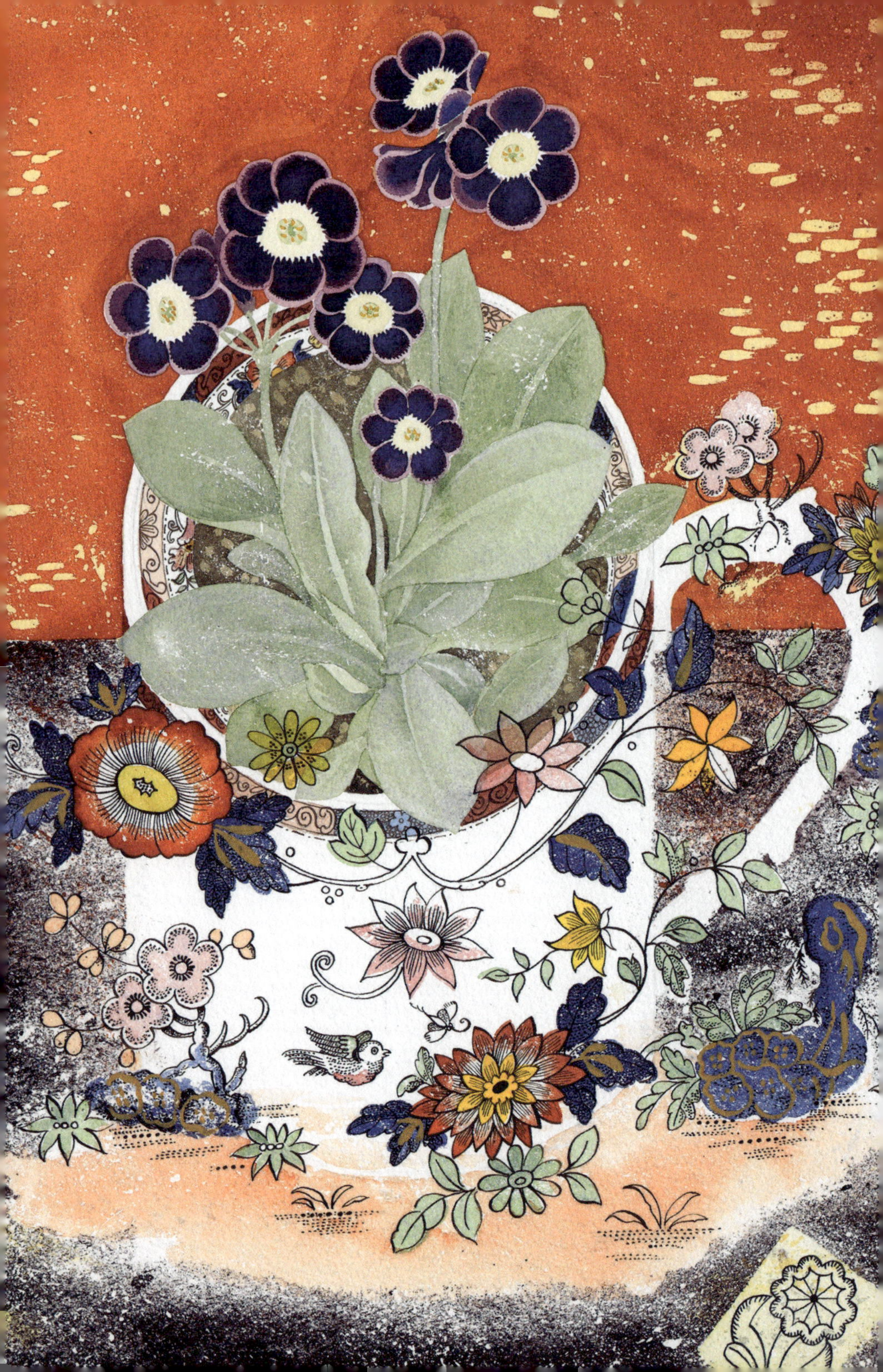

11

AURICULAS

The first time I saw a fancy auricula I fell in love. The ancestors of these sturdy, often brilliantly coloured little primrose relatives grow wild on windswept mountainsides in the European Alps, and, in common with many other alpine plants, have adapted themselves to extremes of climate. Buried under snow all winter, baked by the sun all summer and buffeted by gales, they're tough as old boots but rather more alluring in the garden.

Wild auriculas (*Primula auricula*) have a rosette of smooth, rather rubbery green leaves with toothed edges, and it's the shape of the leaves that won them their name; early botanists gave them the Latin appellation *Auricula ursi*, literally 'little bear ears'. Their canary-yellow flowers are borne in clusters on a robust stem, and it's easy to see why they were commonly called mountain cowslips. Some, although not all, of them also have what looks like a dusting of white flour on their leaves and in a ring around the centre of their flowers. This farina, as it's called, is shared with many other members of the *Primula* family, and is probably another alpine adaptation, which may reduce the damaging effects of the sun's rays at high altitude.

But if wild auriculas are yellow, why do garden auriculas come in such a striking array of colours? The answer has to do with the wonders of hybridization, both natural and human-led. It's thought most likely that our garden varieties originated as a cross between *P. auricula* and *P. hirsuta*, a dinky little wild primula with multi-headed pink (or sometimes white) flowers, that grows in similar conditions in the Alps. The resulting natural hybrid, properly known by its Latin name, *P. × pubescens*, seems to have been what was first collected from the wild by keen-eyed gardeners, probably during the sixteenth century.

By the early seventeenth century a number of garden hybrids were being grown, and as Brenda Hyatt mentions in her excellent monograph, in the 1630s they were among the rare flowers displayed and bred by Charles I's head horticulturist, John Tradescant, in his famous garden at Vauxhall. Soon there were numerous colours to choose from, and in his *Garden Book* of 1659, Sir Thomas Hanmer wrote: 'We have whites, yellows of all sorts, haire colours, oranges, cherry colours, crimson and other reds, violets, purples, murreys, tawneys, olives, cinammon colours, ash colour, dunns, and whatnot…'

Page 84: *Auricula Faliraki*
Watercolour, 2020

Yellow Auricula in Floral Cup
Watercolour, 2022

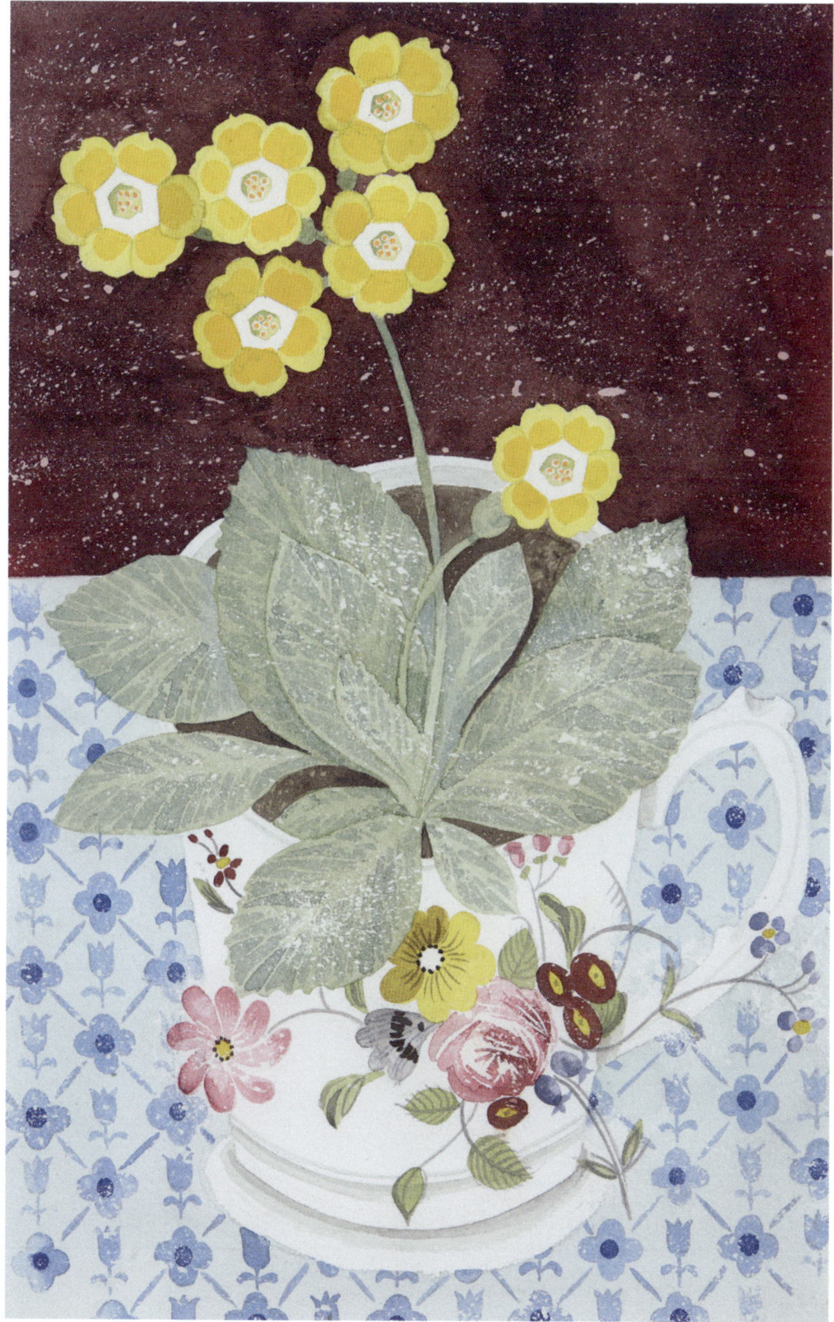

Continued hybridization soon produced double flowers, and in about 1750 came the first 'edged' auriculas, which had a delicate ring of green, grey or white farina around the edges of their flowers. These quickly became so popular that the older varieties fell out of fashion, while the edged flowers became known abroad as English auriculas.

In common with so many other plant introductions, fancy auriculas began their garden career as rare exotics, grown and prized by the wealthiest echelons of society. But as they were more widely grown their popularity spread from the rich to the middle classes and, by the mid-eighteenth century, to working-class gardeners, especially in the North of England, where they were grown as competitively as gooseberries and onions. As Hyatt writes, 'No efforts were spared by rival growers, who displayed their plants in miniature lamplit "theatres". Many of these ornate theatres had backgrounds of black velvet, with mirrors placed in position to show the flowers from every conceivable angle. Others had colourful painted backgrounds of mountain scenes. It is from these theatres that the term "stage auriculas" comes.'

By the 1830s auriculas were one of the flowers to feature in 'florists' feasts' – amateur competitions generally held in local pubs. These distant ancestors of the Chelsea Flower Show were advertised with a copper kettle hung outside, which formed the main prize, but by the 1860s they had been superseded by larger, more formal events, culminating in the formation of the National Auricula Society, which held its first show in 1873. Although auriculas' popularity declined in the twentieth century, the Society soldiered on, and is still going

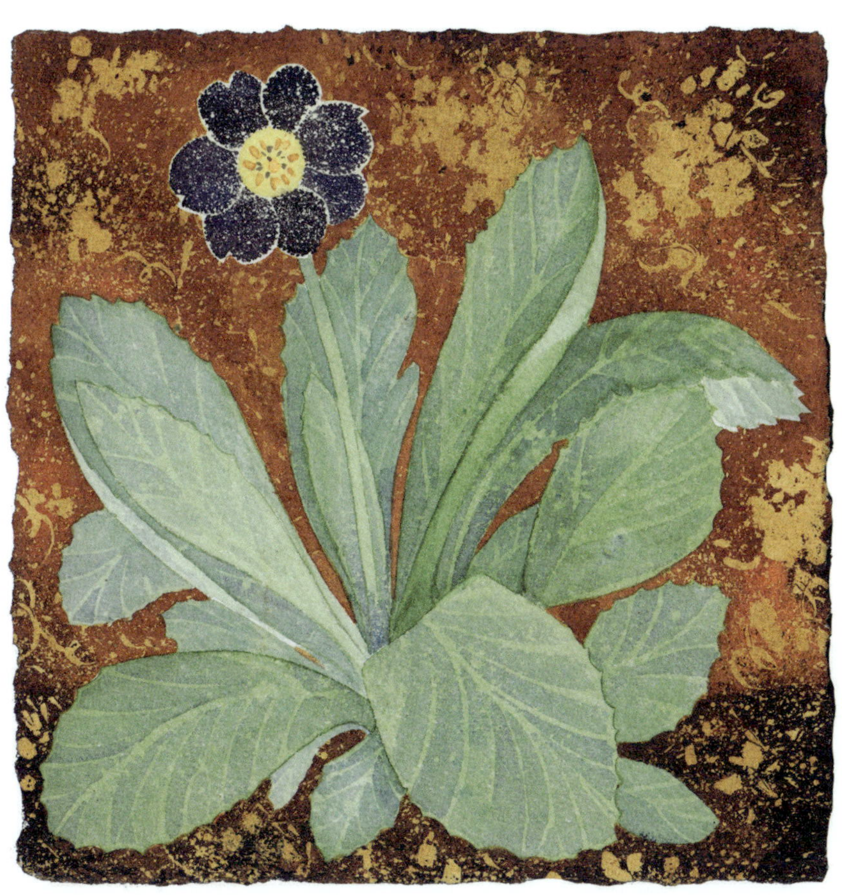

strong today, now split into three regional 'sections' (Northern, Southern, and Midland & West). Their shows, along with those of the Royal Caledonian Horticultural Society north of the Scottish border, are well worth seeking out. A bench of fancy auriculas, with their jewelled flowers on almost comically sturdy stems, is a magical sight, and you might well fall in love with them too.

Overleaf: *Auricula Tapestry*
Watercolour, 2020

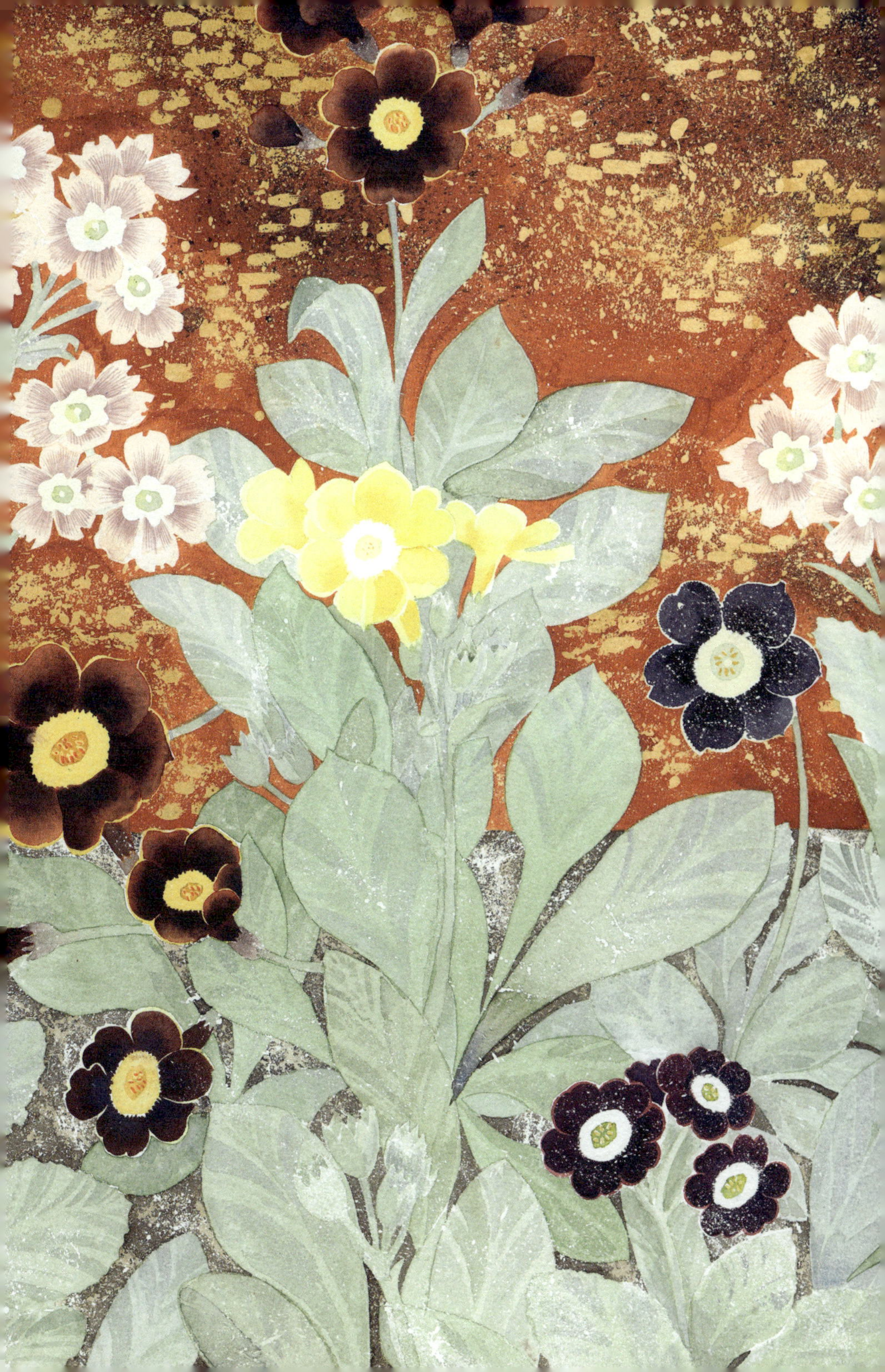

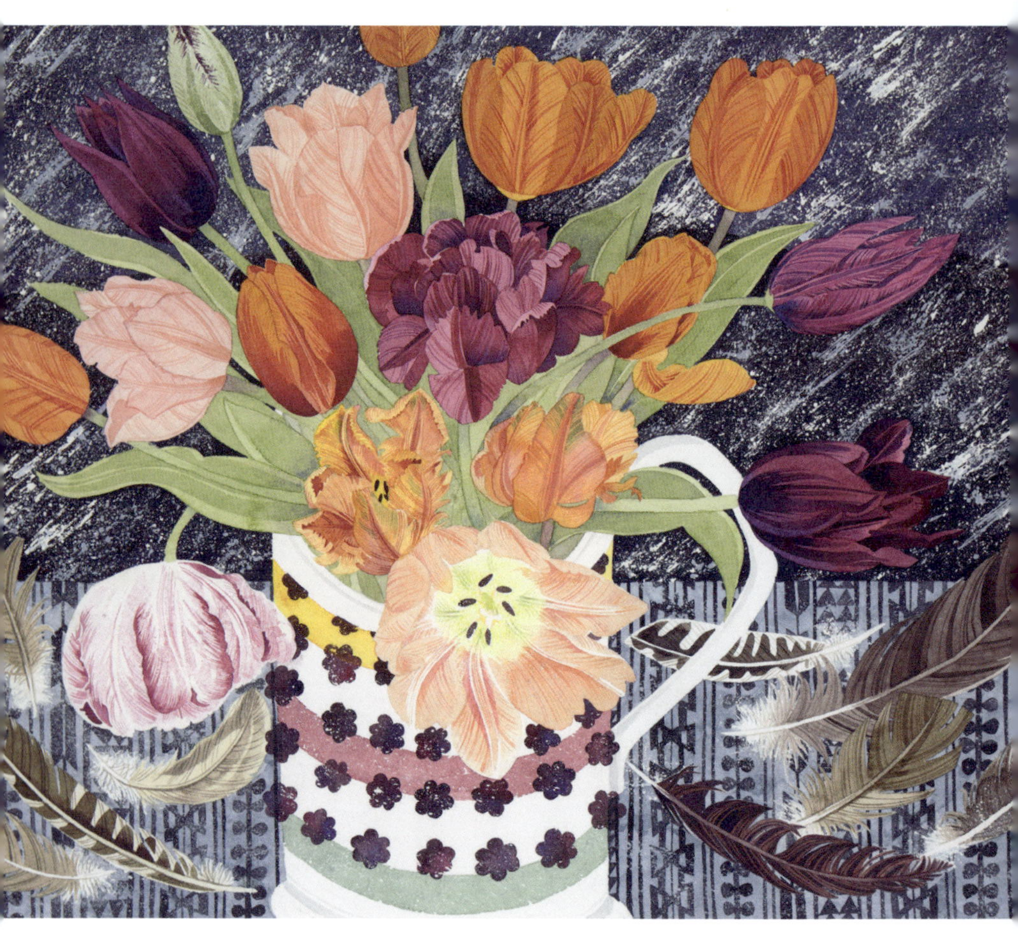

Page 95: *Tulips in Lustreware Jug*
Watercolour, 2023

12

TULIPS

There are so many wild species and cultivated varieties of tulip that you could easily write an entire book about them – as did the esteemed garden writer Anna Pavord. Her 1999 bestseller, entitled simply *The Tulip*, told their story in fascinating detail and is still a great read today. I won't attempt to replicate her feat in this short chapter, even if I had her talent or knowledge, nor does there seem to be any sense in repeating the often-told tale of how, during a brief period of speculative madness between 1634 and 1637, tulips became such hotly traded commodities in Holland that entire fortunes were made and lost. So what does that leave to say about them?

Plenty. There are currently more than eight thousand different tulips on the internationally recognized list maintained by the Koninklijke Algemeene Vereeniging voor Bloembollencultuur (otherwise known as the Dutch Royal General Association for Bulb Culture), so there is no shortage of varieties to choose from. But what explains their lasting popularity? After all, they're rather simple-looking flowers compared with many others, and you wouldn't grow many of them for their foliage alone. They rarely produce more than a single flower per bulb, and those are only briefly in bloom. Yet maybe it's precisely their simplicity that appeals, with all the thousands of cultivars offering so many variations on a simple theme, a bit like a horticultural version of Beethoven's *Diabelli Variations*.

I love most kinds of tulip, whether plain or parrot, lily or Rembrandt, broken or bizarre, but nothing beats seeing natural species growing in their native habitat, which often comes as a surprise when you've only seen them in a garden before. There are around seventy-five wild species, growing mainly in upland regions along a northern latitude of 40 degrees, from Greece and Turkey in the west to the Tian Shan mountains that straddle the borders of China, Kazakhstan and Kyrgyzstan.

The majority of tulips we grow at home are the result of centuries of hybridization between various wild species, especially *Tulipa suaveolens*, which originates in the steppes of Ukraine and Crimea, but some individual species make excellent garden plants as well. Most garden hybrids have to be treated as annuals and thrown away after flowering, since their vigour and quality go rapidly downhill – a wasteful exercise in an era of strained resources. Apart from their inherent beauty, species tulips have the advantage of coming back year after year, making them perfect for naturalizing in borders and wild-flower meadows. Among the easiest are *T. saxatilis*, which has pale pink flowers and a golden centre, the scarlet *T. sprengeri*, from

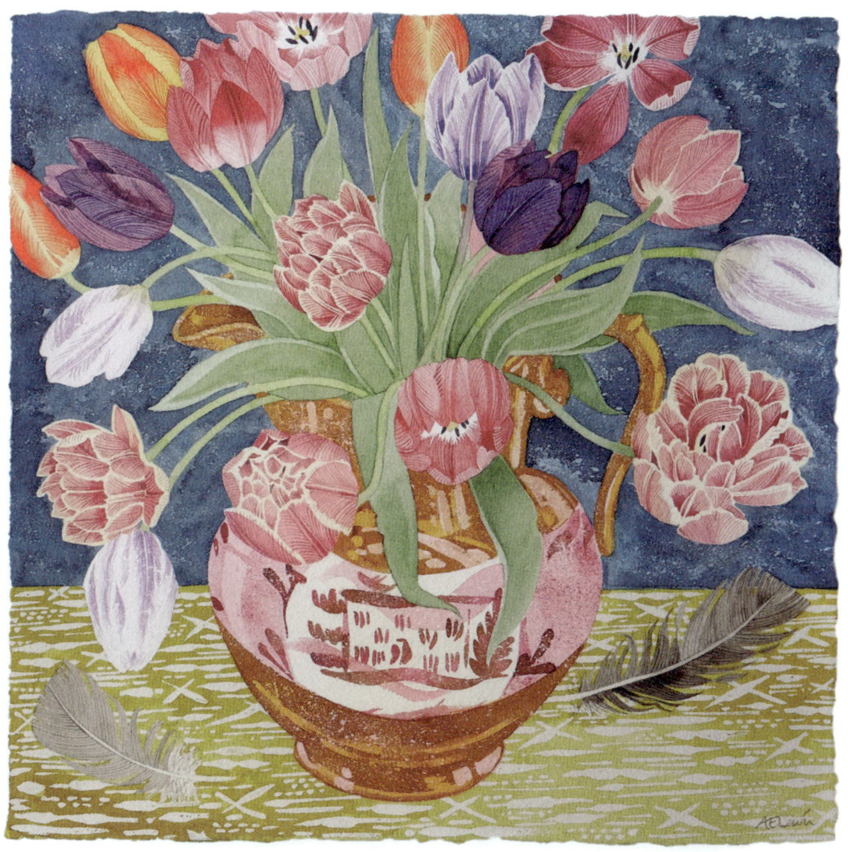

the Turkish Black Sea coast, and *T. turkestanica*, with striking white flowers around yellow centres.

My own favourite is *T. orphanidea* (previously known as *T. whittallii*), whose petals range between red and burnt orange, with sharply pointed tips. It's a beautiful little tulip, found in Turkey, Crete and mainland Greece, which is where I came across it a few years ago on a visit to the northern mountains of the Peloponnese in early May. We had been driving south from Kalavrita, on the way to the spectacular source of the River Aroanios, which springs fully formed out of the ground in a beautiful forest of plane trees near the village of Planitero. A few kilometres off the main road, however, our map showed the site of the ancient Greek temple of Artemis Hemera, so we decided to make a detour and have a look. The temple, which dates back to the ninth century BC, sits on the side of a steeply wooded hill above the remains of the ancient city of Lousoi, with views across the plain of Soudena to the mountains of Achaea. Like so many smaller Greek sites, it is magically set and was completely deserted, so that we were the only visitors.

As the narrow road looped back down to the plain, it ran between roughly ploughed fields of soil that seemed to be composed largely of stones, but as we drove we saw points of bright colour, and (not for the first time) skidded to a halt. Clambering up a bank to the level of the field, we found mostly bare earth, but bare earth scattered with beautiful little tulips with sharply pointed flowers, scarlet outside and orange-red inside, with black blotches at their centres and six prominent stamens dusted with bright yellow pollen: *Tulipa orphanidea.* They weren't, perhaps, as attractively planted as they might have been by a gardener, but somehow it was far more thrilling to come across them by chance, thriving like weeds in their natural habitat. And now, whenever I see them in a garden, I remember how we stumbled across them at the side of the road in Greece.

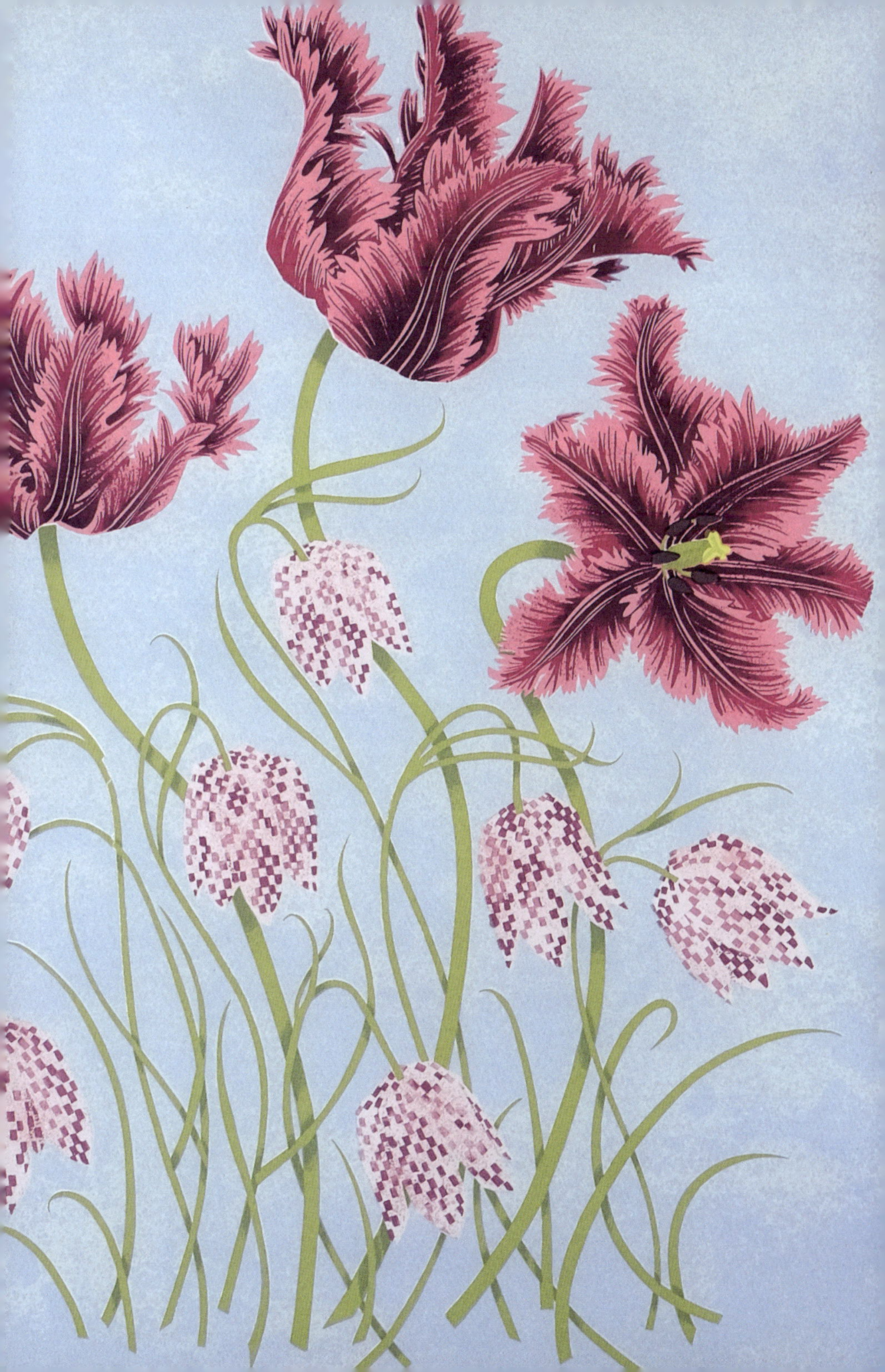

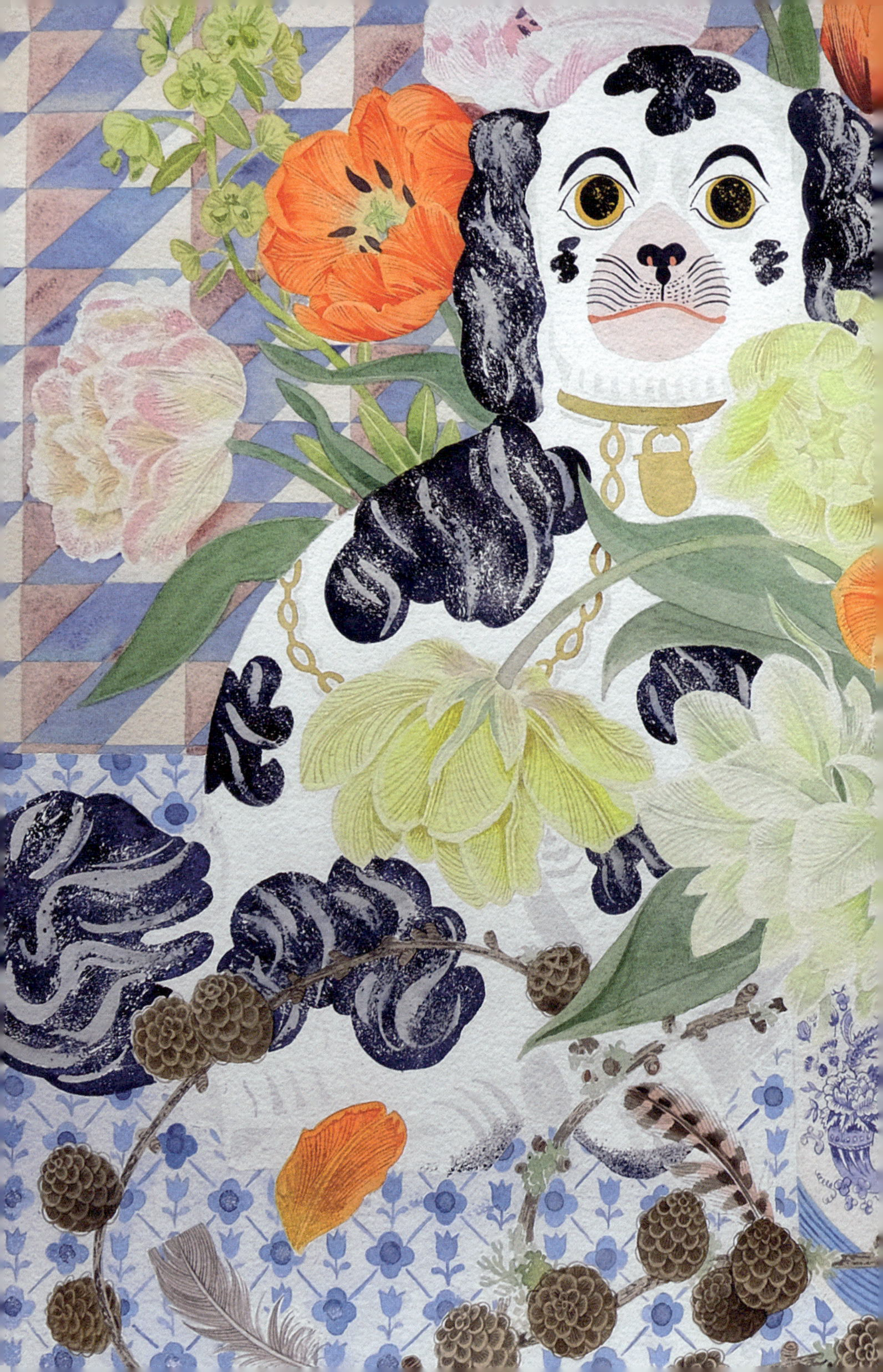

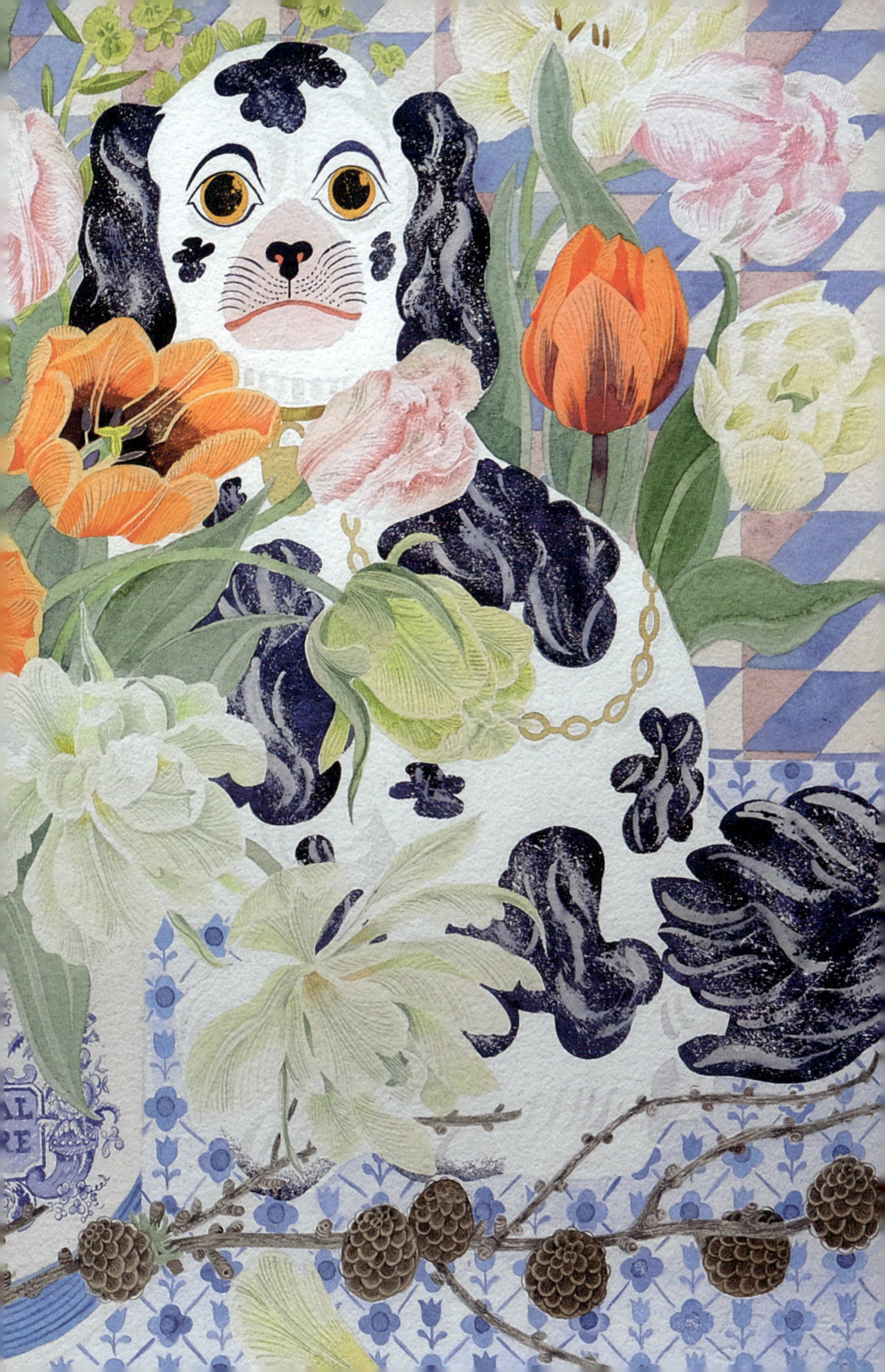

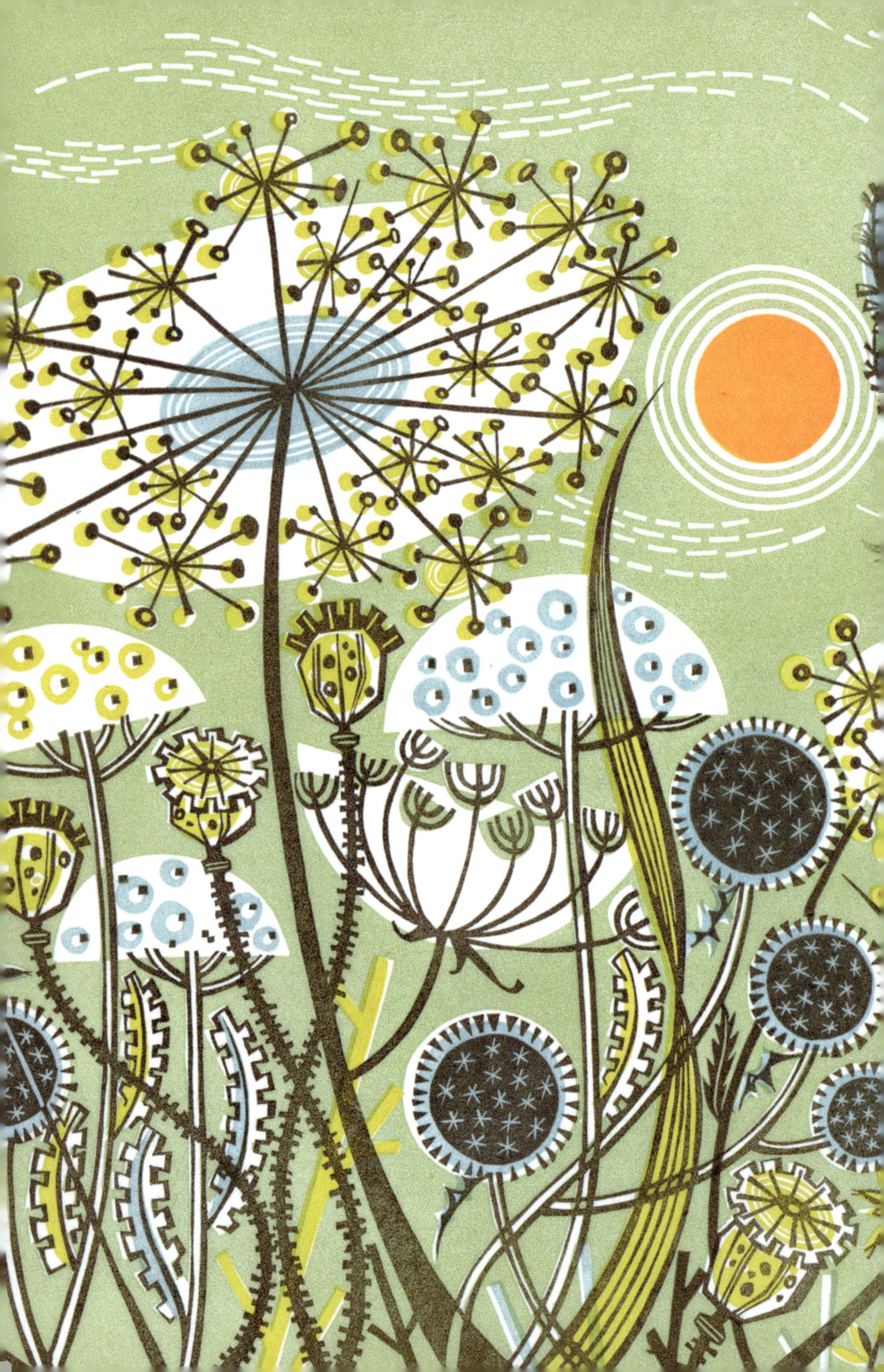

13

ECHINOPS

Tall, striking and resilient, it's not surprising that globe thistles have become such a popular garden flower. As do other thistles, they develop deep tap roots that help them to shrug off droughts, and they will grow happily in the poorest soil. They're also pollinator magnets, and some species have true-blue flowers, making them a useful addition to the midsummer colour palette.

Globe thistles (or *Echinops*, to give them their botanical name) belong to the daisy family, along with sunflowers, dahlias and thistles, and, as with other members of the Asteraceae, their flowerheads are made up of hundreds of individual florets. In globe thistles, however, unlike in sunflowers and daisies, these florets are arranged in a globular shape, which is what gives them their common name.

Green Meadow (detail)

Linocut, 2009

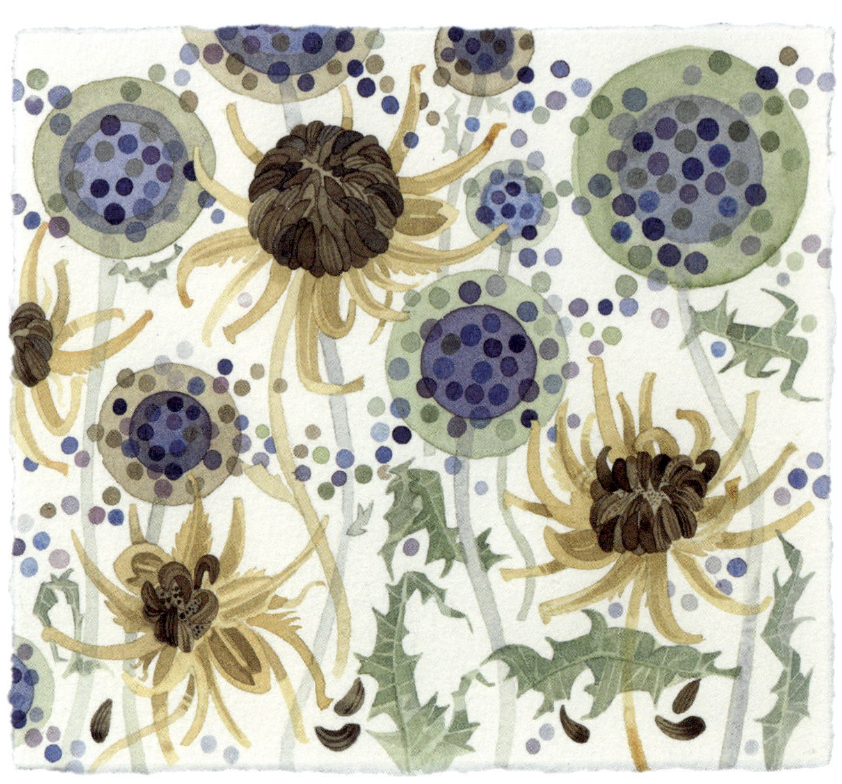

Echinops and Calendula

Watercolour, 2020

Echinops and Primula

Watercolour, 2020

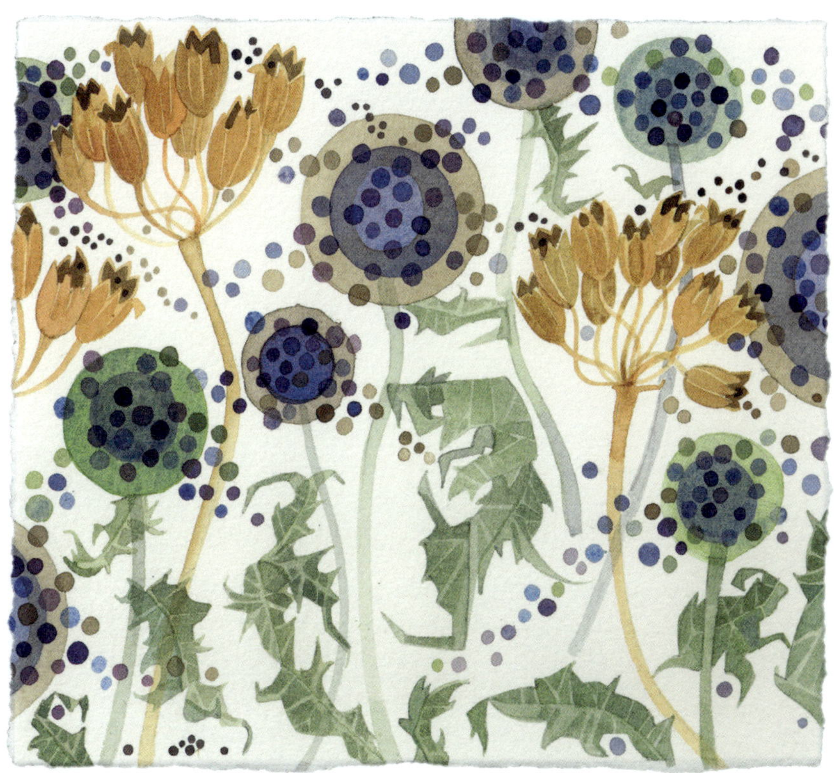

Their botanical moniker (given to them by Carl Linnaeus in 1753) is just as descriptive: *ekhinos* is the ancient Greek for hedgehog, as well as for other spiny objects such as sea urchins and sweet chestnuts.

There are about 120 species in the genus, but we grow only a handful of them in our gardens. Perhaps the earliest to arrive was the white-flowered *E. sphaerocephalus*, which grows wild across a huge area from central Europe through Russia and Siberia as far east as Xinjiang, China. The plant historian Maggie Campbell-Culver dates its arrival in Britain to 1542, followed in 1570 by *E. ritro*, which is still the most commonly grown species today, especially in its darker blue form, *E. ritro* 'Veitch's Blue'.

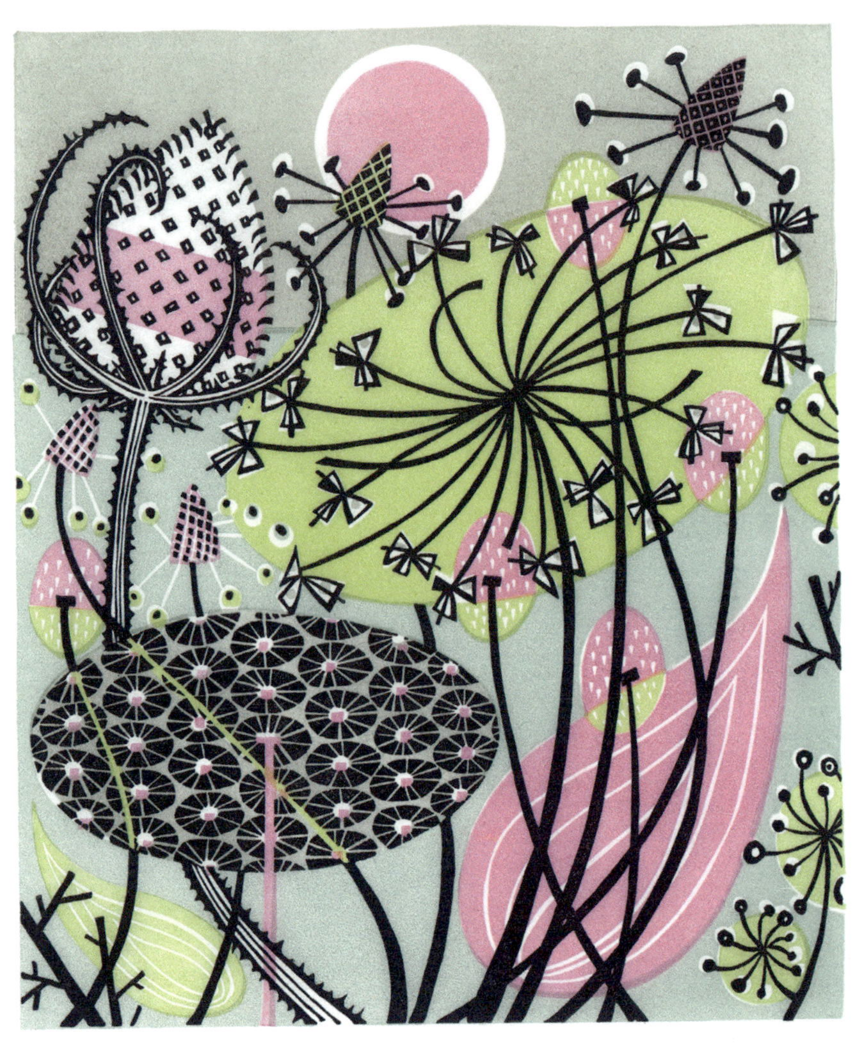

Alliums

Linocut, 2007

14

ALLIUMS

How well do you know your onions? I guess that depends what you mean by onion, but if we include all the members of the *Allium* genus (to which onions, garlic, leeks, chives and shallots all belong, as well as the many decorative alliums we grow in our gardens), we're talking anything from three hundred to nearly one thousand species. That's a lot of onions.

Alliums are generally pretty easy to identify, since most of them have a similar appearance: growing from a smaller or larger bulb, they tend to have long, narrow leaves and a single, leafless flower stem, with a starburst of small flowers at the top, usually pink, purple or white. In the wild they're confined to the northern hemisphere, although there may be a couple of examples in Chile and Brazil, and there's a single, puzzling species in South Africa, which no one can quite account for.

One thing they all have in common, though, is their smell. The onioniness of onions is not to everybody's taste, and that's the point: their penetrating fragrance evolved as a defence against browsing animals, most of which avoid alliums like the plague. It might also explain their relatively late adoption as an ornamental (rather than a culinary) plant. After all, it's one thing to swish through drifts of roses and rosemary, but quite another to brush through a bed of alliums and emerge smelling like an onion bhaji.

One of the best-documented early botanic gardens was created in the south London suburb of Vauxhall during the seventeenth century by John Tradescant the Elder and his imaginatively named son, John Tradescant the Younger. Their house, which became known as The Ark for its all-encompassing collections of curiosities, had an equally celebrated garden, in which the Tradescants grew hundreds of exotic plants, many of them – such as tradescantia – for the first time in Britain. Yet their catalogue from 1656 lists only four alliums: garlic (*A. sativum*), ramsons (*A. ursinum*), crow garlic (*A. vineale*) and something called 'Great Turkey Garlic', which may or may not be what we now call elephant garlic (*A. ampeloprasum* var. *ampeloprasum*).

Two hundred years later, alliums were hardly any more popular. In *The Wild Garden*, published in 1870, William Robinson – arguably the most influential garden writer of the nineteenth century – was willing to admit that 'Some of the species are very beautiful, so much so as to claim for them a place in gardens notwithstanding their disagreeable odour.' But he went on to caution, 'It is in the wild garden only, however, that this family

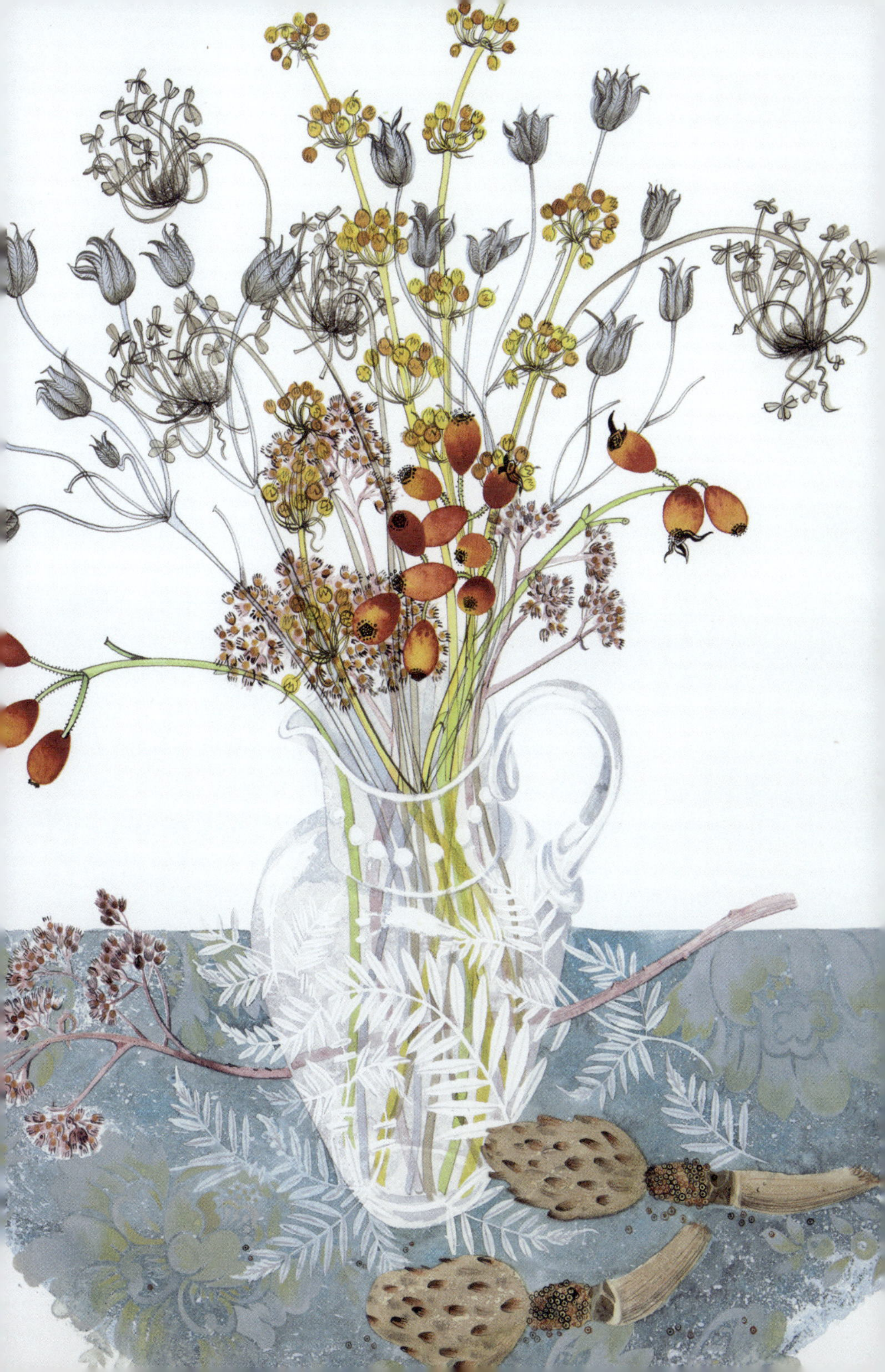

can find a fitting home,' and he recommended only two species: *A. neapolitanum* and *A. ciliatum* (since renamed *A. jubatum*).

Things picked up for alliums in the twentieth century, perhaps because gardeners became less anxious about 'disagreeable odours', but also because alliums generally don't require great horticultural expertise. By the mid-1960s they were finally starting to become fashionable, if a popular practical gardening book of the time is anything to go by. *The Bulb Book* was first published in 1964, and in it the Swiss naturalist Paul Schauenberg enthused that, 'Having experimented with a dozen or so species, I am convinced that alliums are some of the easiest bulbs to grow, and some of the finest. Any reader looking for something out of the ordinary will be absolutely delighted with them and his [sic] friends will be greatly impressed, for their multi-coloured balls of flowers are unfailingly spectacular.'

Schauenberg lists twenty-seven species to choose from, including many that are still popular today. *A. aflatunense*, which grows wild in Kazakhstan and Kyrgyzstan, has sturdy lilac flowers, but over the years it has been completely muddled up in cultivation with *A. hollandicum*, which originates in Iran and has taller, darker flowers. The problem seems to date back to 1948, when the Dutch bulb company Van Tubergen started selling *A. hollandicum* under the name *A. aflatunense*, and you're much more likely to end up with the former than with the latter, whatever the label says; some sellers hedge their bets by calling it *A. aflatunense hollandicum*. The most popular cultivar, *A. hollandicum* 'Purple Sensation', dates back to 1963, and was selected by a leading Dutch allium specialist, Jan Bijl, who was also behind the enormous *A.* 'Globemaster'.

A. cernuum, also known as the nodding onion, is found in widely scattered populations across North America, and, as its common name suggests, its flowers dangle in an elegant cluster from the stem. Although white varieties are available, most are purplish-pink, with protruding yellow stamens making an attractive colour contrast. Small to middling in height, they're good for meadow gardens, not least because (in common with other alliums) they're relatively deer- and rabbit-resistant.

Arguably the most striking species of all, *A. cristophii*, has huge spherical flowers – or, more accurately, clusters of flowers called umbels – with starry individual florets that have shiny, almost metallic-violet petals. They have an interesting history. Native to the Kopet-Dag mountain range, which runs for hundreds of kilometres along the north-eastern border of Iran, their bulbs were first collected by an adventurous Russian diplomat called (deep breath) Baron Clément Augustus Gregory Peter Louis de Bode. De Bode, who described his adventures in his memoir *Travels in Luristan and Arabistan* (1845), sent some bulbs to the director of the St Petersburg Botanic Gardens, and the species was initially named *A. bodeanum* after him.

In 1884, however, a Russian scientist renamed it *A. cristophii*, after one Dr Cristoph, who seems to have visited the Kopet-Dag mountains himself and collected further bulbs. Like *A. aflatunense* (or, more accurately, *A. hollandicum*), it was first sold commercially by Van Tubergen in the early twentieth century – but they misnamed it *A. albopilosum*, adding another layer of obfuscation. If you're confused by now, join the club – and note that *A. cristophii* is so often misspelled *A. christophii* (with an h) that it's been suggested the original spelling be dropped altogether.

Perhaps even more striking in form is *A. schubertii*, whose spidery flowers are the closest horticultural thing to fireworks going off, with a starburst of tiny florets on stems of varied lengths, forming an openwork globe up to 30cm across. They're not especially tall, but they make a spectacular show, and are much liked by florists. They come from Turkey and the Levant, and were named after the early nineteenth-century German naturalist Gotthilf Heinrich von Schubert, who was admired by Goethe and, in 1814, published *The Symbolism of Dreams*, which later influenced both Freud and Jung.

At the opposite end of the scale, the drumstick allium, *A. sphaer-ocephalon*, is a pretty little European allium with plum-shaped, plum-coloured flowers on longish stems. It looks best en masse and flowers later than most of its relatives; in the right conditions it will multiply quickly – although not as quickly as my bête noire: *A. triquetrum*. Often known as the three-cornered leek and sometimes cooked as an alternative to wild garlic (*A. ursinum*), these small, white-flowered alliums from southern Europe are not unattractive, but in my experience they're horribly invasive. They set seed readily, but also shed tiny bulbs underground, each of which will turn into a new plant. One big garden supplier puts a positive spin on their nasty habits by recommending them as a good choice to suppress weeds and cover bare ground, but be careful what you wish for…

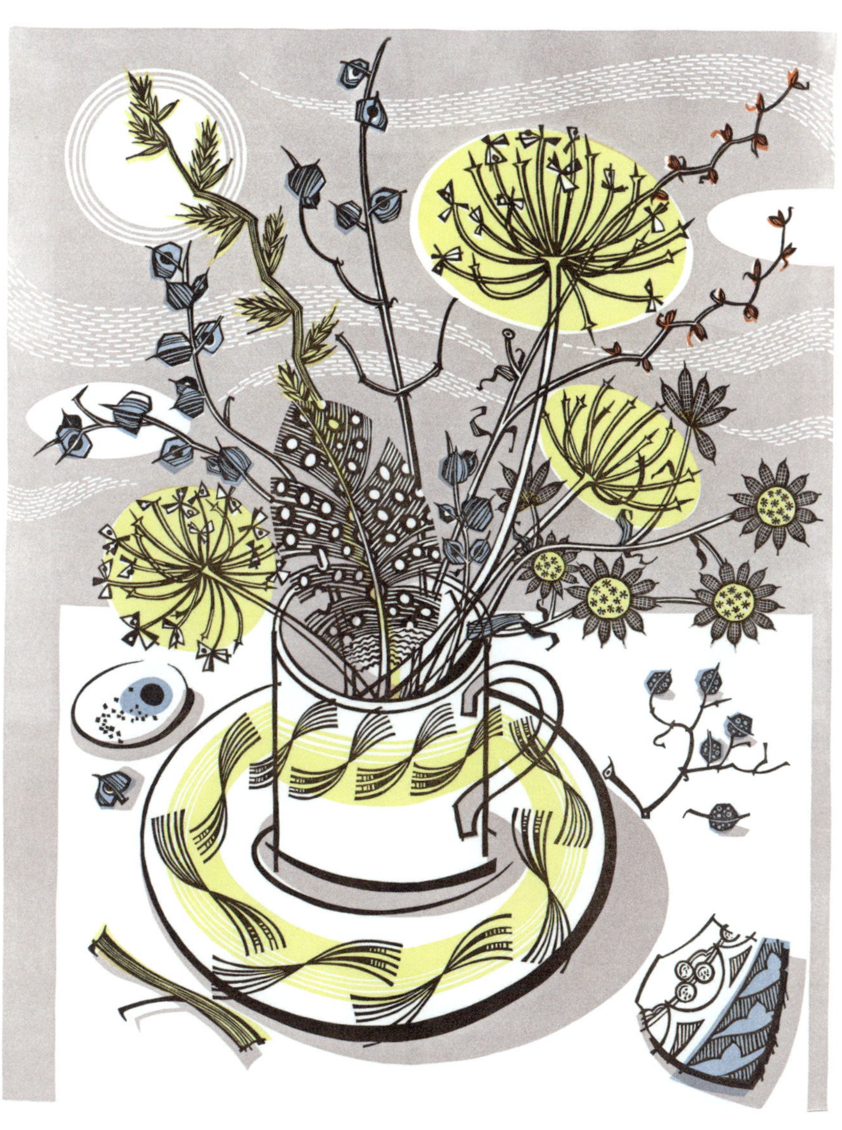

The Moonlit Cup
Linocut, 2008

Overleaf: *Alliums and Fennel*
Lithograph, 2008

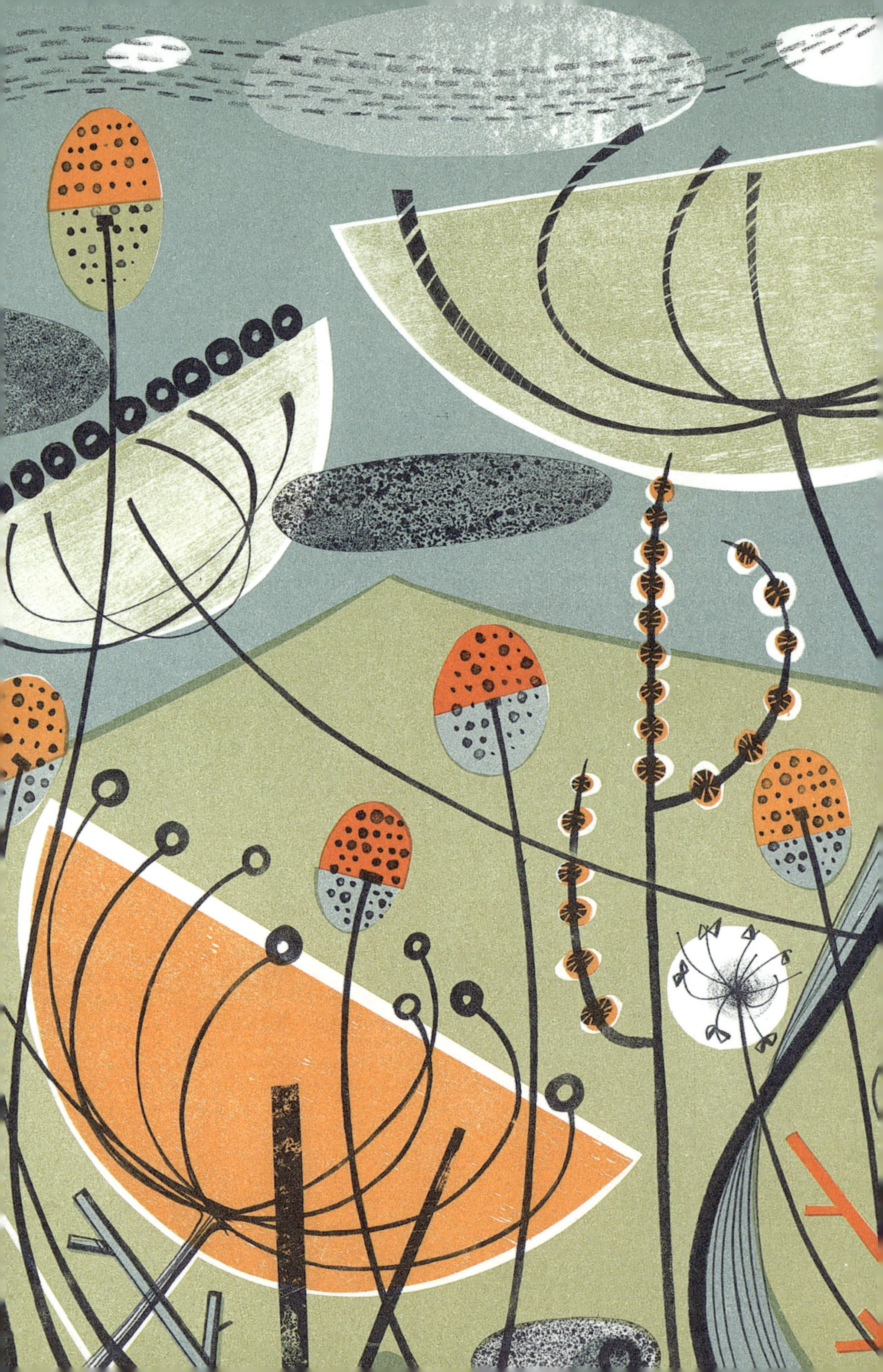

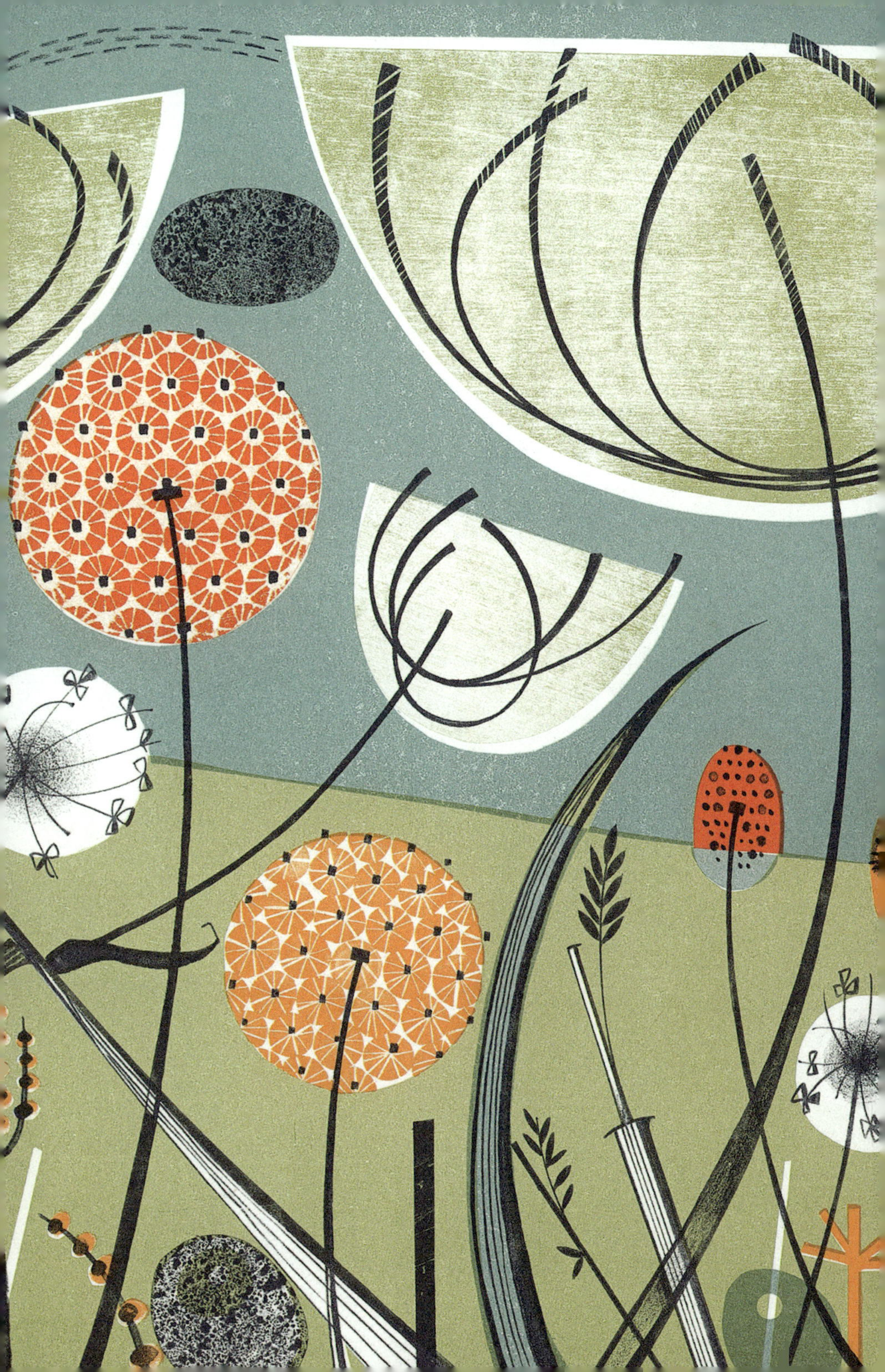

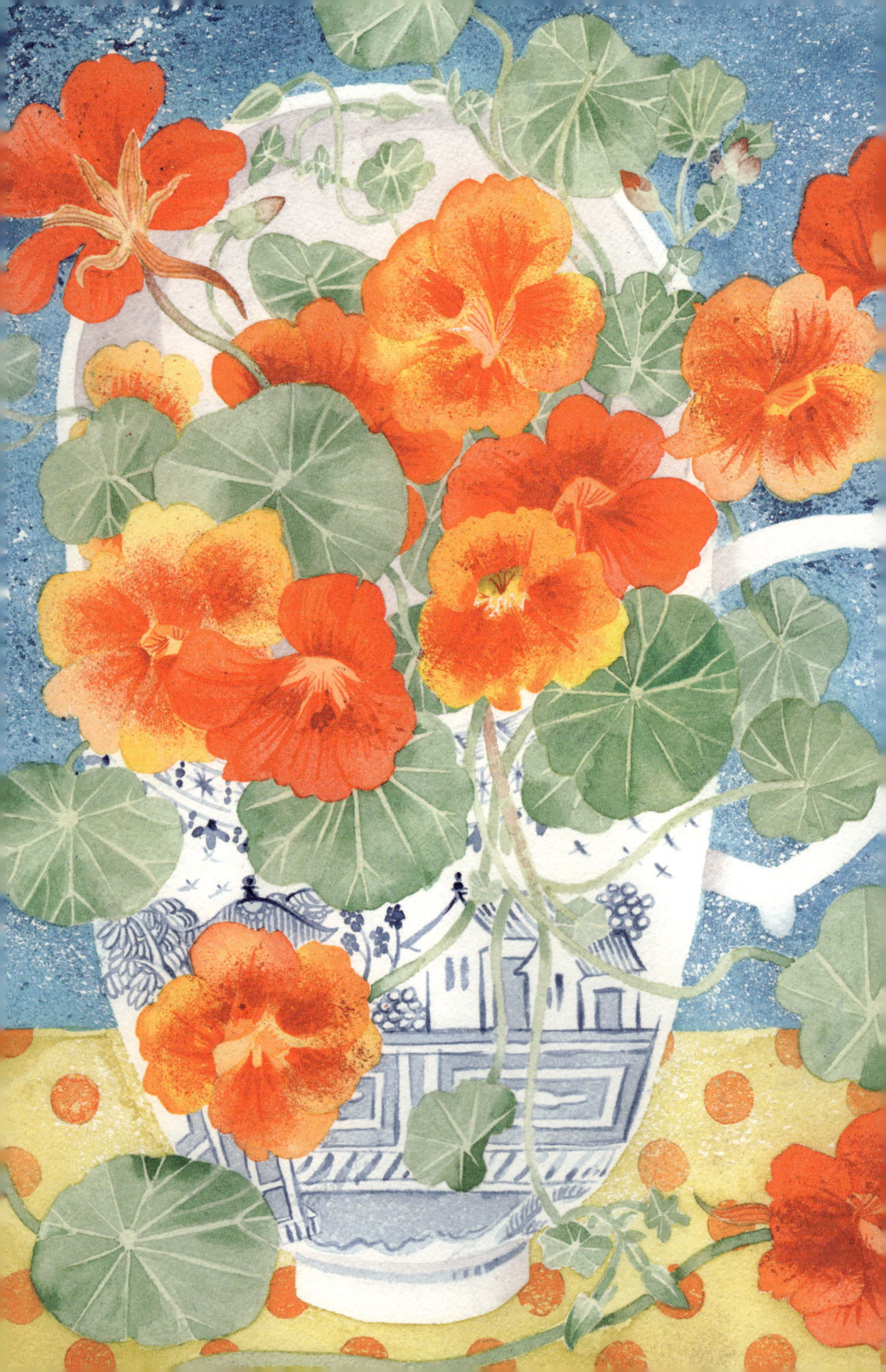

15

NASTURTIUMS

Fetter Lane runs roughly north to south between High Holborn and Fleet Street in the City of London. Today it's something of a backwater, lined with modern office blocks, although a few Victorian and Edwardian buildings survive towards its northern end, and its southern stretch is dominated by the former Public Records Office, opened in 1859 as 'the strong box of the Empire'. In the fourteenth century the street was called Faytor Lane, which seems to derive from the Old French word for lawyer, *faitor*, of which there were (and are) plenty nearby. It may be hard to imagine now, in the intensely urban environment of contemporary EC4, but back in the sixteenth century Fetter Lane still had some of the feeling of a country lane, as illustrated on the famous Agas Map of 1561. Houses are built along most of its length, but behind them are

Late Summer Nasturtiums (detail)
Watercolour, 2023

small fields (or large gardens), and north of High Holborn is still open countryside.

Fetter Lane is where the physician, gardener and herbalist John Gerard lived between the late sixteenth century and his death in 1612, in a house owned by William Cecil, Lord Burghley, whose gardens Gerard superintended in London and Hertfordshire. Gerard's house (which may lie somewhere underneath the Public Records Office) had a garden of its own, and it was there that he built up a remarkable stock of plants from around the world, many of them newly arrived in England. In 1596 Gerard published the world's first garden catalogue, listing the plants in his collection, which by then included more than a thousand rare and unusual specimens, many of which had never been grown in Britain before. Among them were potatoes (then still a novelty), saffron, frankincense, an almond tree – and nasturtiums, whose seeds had been sent to him by Jean Robin, gardener to the kings of France.

Gerard's nasturtium wasn't quite the same as the nasturtiums most of us grow today, although its yellow flowers with orange spots look pretty similar. His was the smaller *Tropaeolum minus*, which originated in the mountains of Ecuador and Peru, and had been brought to Europe as recently as the 1560s by Spanish conquistadors. The vigorous garden nasturtium that we are familiar with, which ramps about all over the place and has orange, yellow or sometimes red flowers, is mainly derived from another Peruvian species, *T. majus*, which arrived in Europe in 1684 courtesy of a Dutch monk called Pater Beverning. He gave some seeds to the botanic garden of the University of Leiden, the Hortus Botanicus, and they spread gradually from there – although the plants we grow today seem to be a nineteenth-century hybrid between *T. majus* and another South American species, *T. peltophorum*, as well as Gerard's original *T. minus*. All very confusing.

Nasturtiums are edible plants, and their leaves contain high levels of vitamin C, which made them popular with scurvy-ridden sailors. Some species have tuberous roots, which are also edible, and *T. tuberosum* (also known as mashua) was an important food plant for the Inca; its pink tubers have a peppery tang. The common garden nasturtium is a useful pot herb too; its flowers add colour to summer salads, and its buds and seeds can be pickled like capers.

Anyone who has visited Monet's house and garden in Giverny in the right season will remember the nasturtiums that cover huge swathes of ground and clamber across the paths there, but perhaps the most spectacular displays can be seen in Boston, Massachusetts. The Isabella Stewart Gardner Museum is famous for its glass-roofed garden courtyard, designed in the style of a Venetian palazzo. Every June, nasturtiums are grown from seed in the museum's own nurseries, transferred to a greenhouse before the first frosts, then trained along wires over the winter; any flowers are removed by hand each day. In early April the plants are unpinned and carefully carried to the third-floor balconies of the museum, from where they hang down into the courtyard in swags up to six metres long. The flowers last for just a few weeks, but it's a brilliant sight to behold.

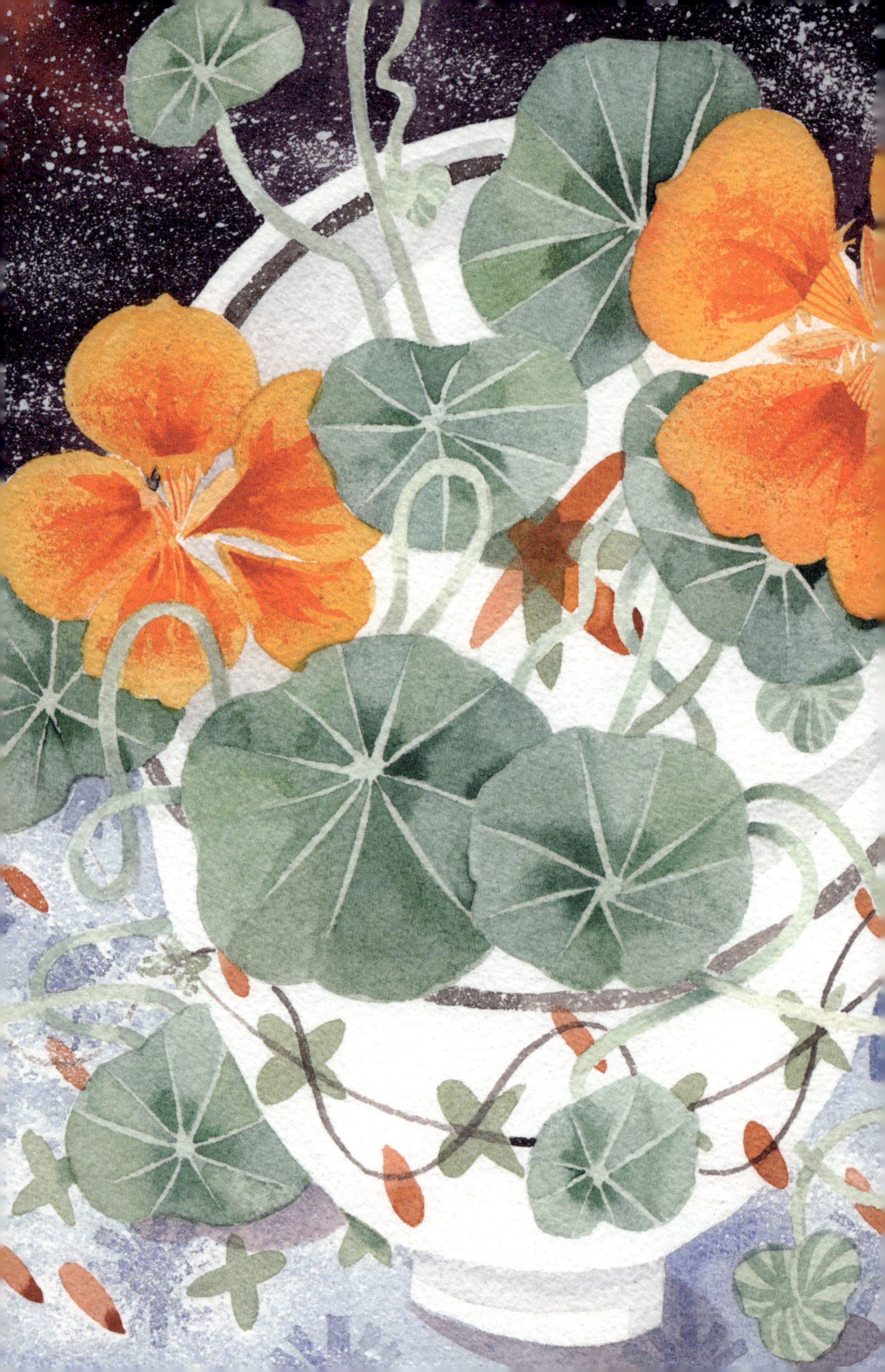

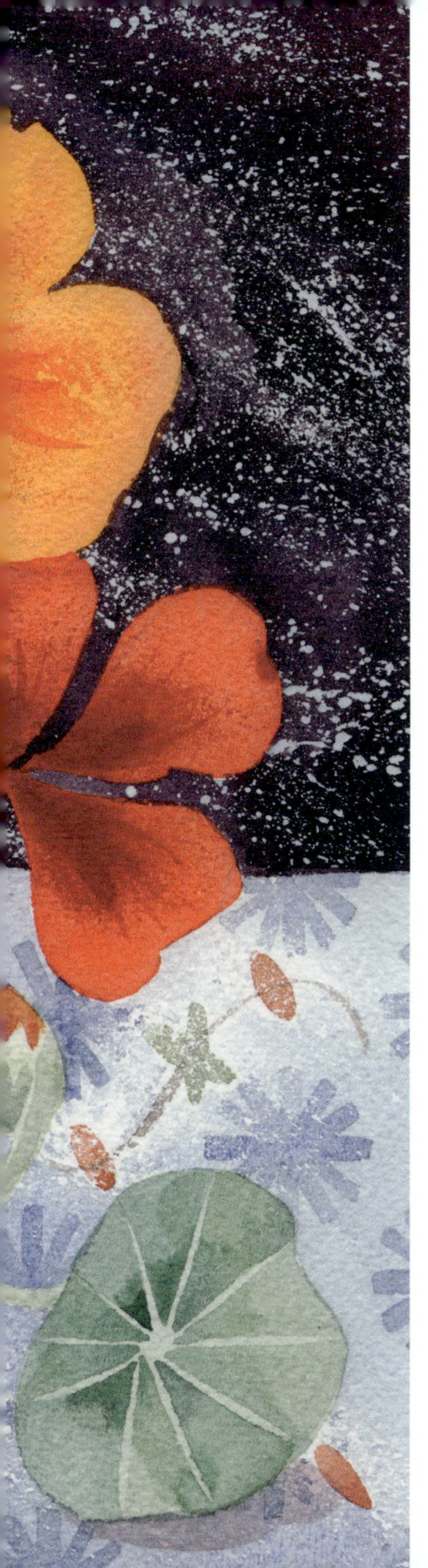

Nasturtiums and Teabowl
Watercolour, 2023

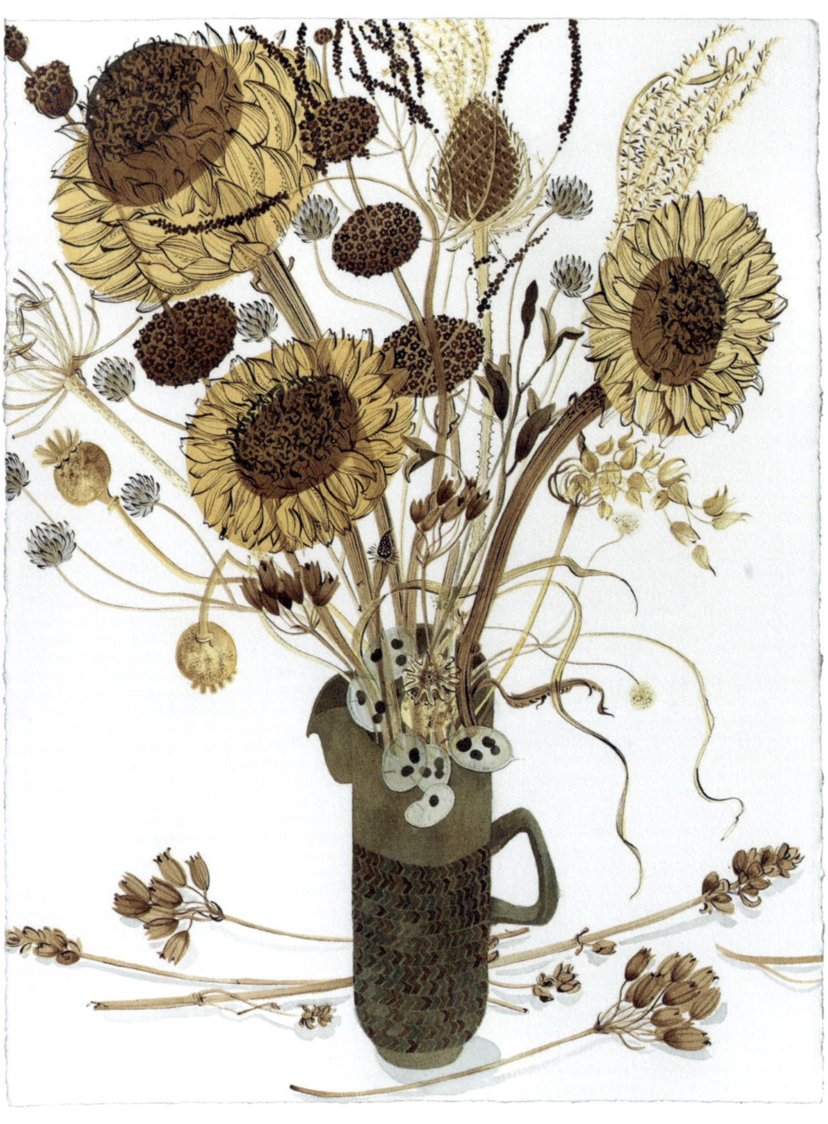

16

ARTICHOKES

A Catholic private school on the outskirts of Chelmsford might
not seem a likely alma mater for British artichokes, but it may well
be the first place where they were grown in this country. New Hall
School has, since 1799, occupied the site and the surviving Tudor
wing of the Palace of Newhall or Beaulieu, built for King Henry
VIII around 1517. It was in the gardens there, if later accounts
are to be believed, that the artichoke – a Dutch import that was
already popular in the courts of France and Italy – was first cultivated for the royal table. Their earliest appearance in print is in a
manuscript dating from 1531, which mentions 'Bringing Archcokks
to the Kings Grace' (that is, into the king's presence); fifty years
later the writer Richard Hakluyt averred that, 'In time of memory
things haue bene brought in that were not here before, as…the
Artichowe in time of king Henry the eight.'

Globe artichokes (to distinguish them from the unrelated Jerusalem kind) are a refined form of the wild thistle, *Cynara cardunculus*, which is better known to most of us as the cardoon. Cardoons grow wild almost right around the Mediterranean, and their fleshy stems were cooked from ancient times. They're tall, handsome, stately plants, with greyish spiny foliage and bright blue thistle flowers, and they make a striking architectural statement at the back of a herbaceous border.

When and where cardoons turned into artichokes we don't really know, but it seems likely that we have the Romans to thank. In a process that may have taken place over hundreds of years, horticulturists took repeated cuttings of wild cardoons that had slightly bigger flowers than the others, and slightly less thorny stems, gradually transforming their appearance – although not so much their genetics – from their uncultivated ancestors. The result was two different forms of the same original plant: one raised for its edible leaves (cardoons) and the other for its edible flowers (artichokes). As with so many other aspects of Roman civilization, it seems to have been the Arabic kingdoms that preserved artichokes after the fall of the Roman empire, introducing them into Spain and Sicily, from where they spread to the rest of Europe; Sicily remains a centre of artichoke production to this day.

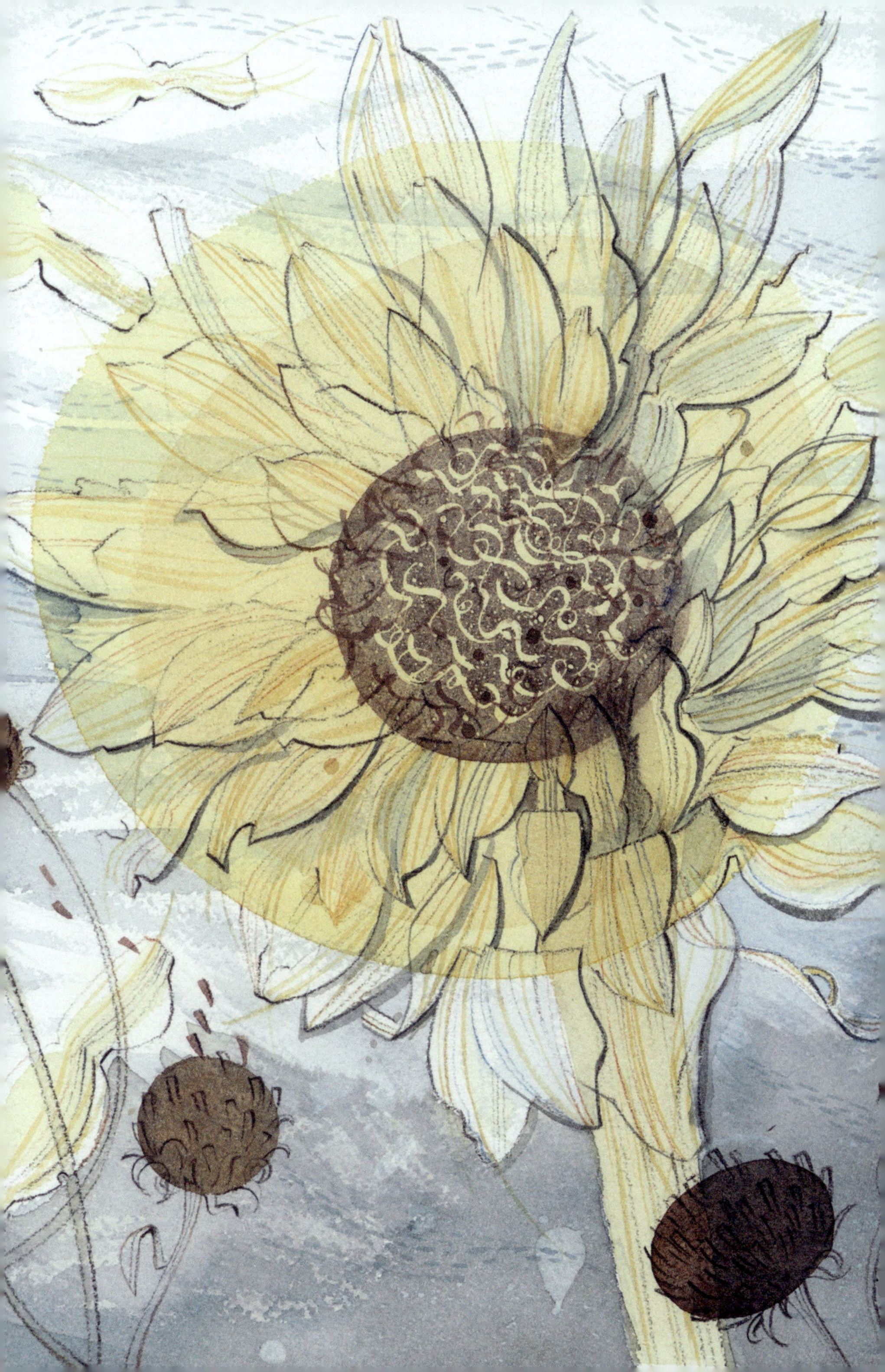

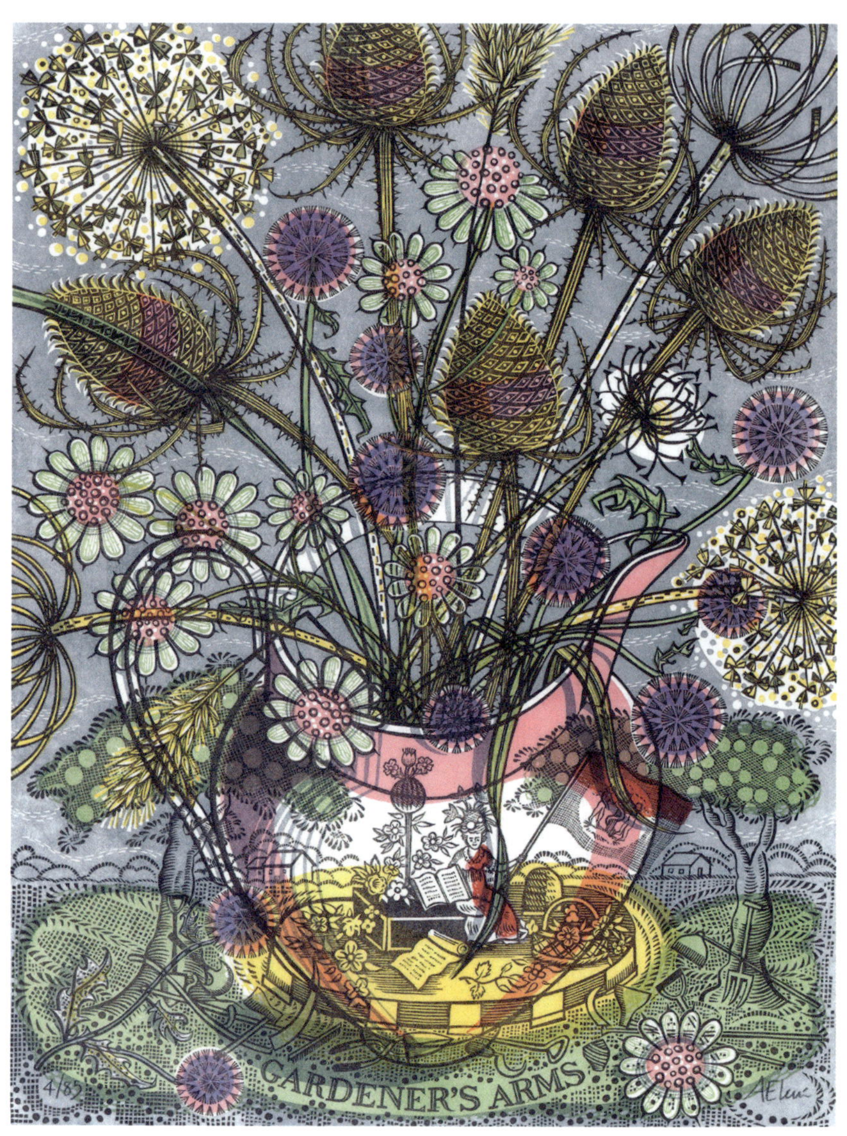
4/85
GARDENER'S ARMS
AELine

17

ASTRANTIA

The Queyras is one of the less well-known regions of the French Alps, running to the Italian border south of Briançon. It's a glorious area for walking and botanizing, with high mountains whose north-facing slopes are cloaked with conifer forests. From a distance these appear to be composed of the standard Norway spruce and silver firs, but closer up it becomes apparent that they are, more unusually, dominated by the European larch (*Larix decidua*). This makes a big difference to the forests' character and ecology, for larch trees are, weirdly, deciduous conifers, which drop their needle-like leaves each winter. They also grow rather further apart than many other conifers, and this, along with their deciduous habit, means that the ground beneath them – although often rocky and precipitously steep – isn't nearly as gloomy and

Page 127: *Lustreware Cup with Astrantia* (detail)
Watercolour, 2020

bare as you'd find in most conifer forests. In fact, larch forests can support a surprisingly rich flora, including anemones, orchids, wintergreen and wild delphiniums.

For me, though, the most surprising sight as we scrambled down through steep forests from the upper slopes was, beneath the trees, great meadows of *Astrantia major*, which I'd seen before only in Chelsea show gardens and herbaceous borders. There's something faintly uncanny about finding a garden favourite growing in such abundance in the wild; they almost always look happier, if somewhat less primped and preened, than their garden-bound compatriots, although no one has ever planted or cherished them.

A. major's pretty, pincushion-shaped flowers are actually in the same family (the Apiaceae) as cow parsley, fennel and carrots, although they lack their characteristic flat-topped 'umbels' of tiny flowers. Instead they have a hemispherical boss of minute white flowerlets, framed by a pale ruff of what look like outer petals but are actually bracts. (Bracts are modified leaves, and they can sometimes be more colourful than the flowers they enclose – bougainvillea is a prime example.) *A. major* is native to most of Europe apart from Scandinavia and Britain, although they were evidently introduced here hundreds of years ago. They were certainly grown by the herbalist and inveterate plant collector John Gerard in the 1590s, from seed that he apparently obtained from Austria. Gerard knew the plant as black masterwort, a common name it still sometimes goes by, along with the charming 'Hattie's pincushion'.

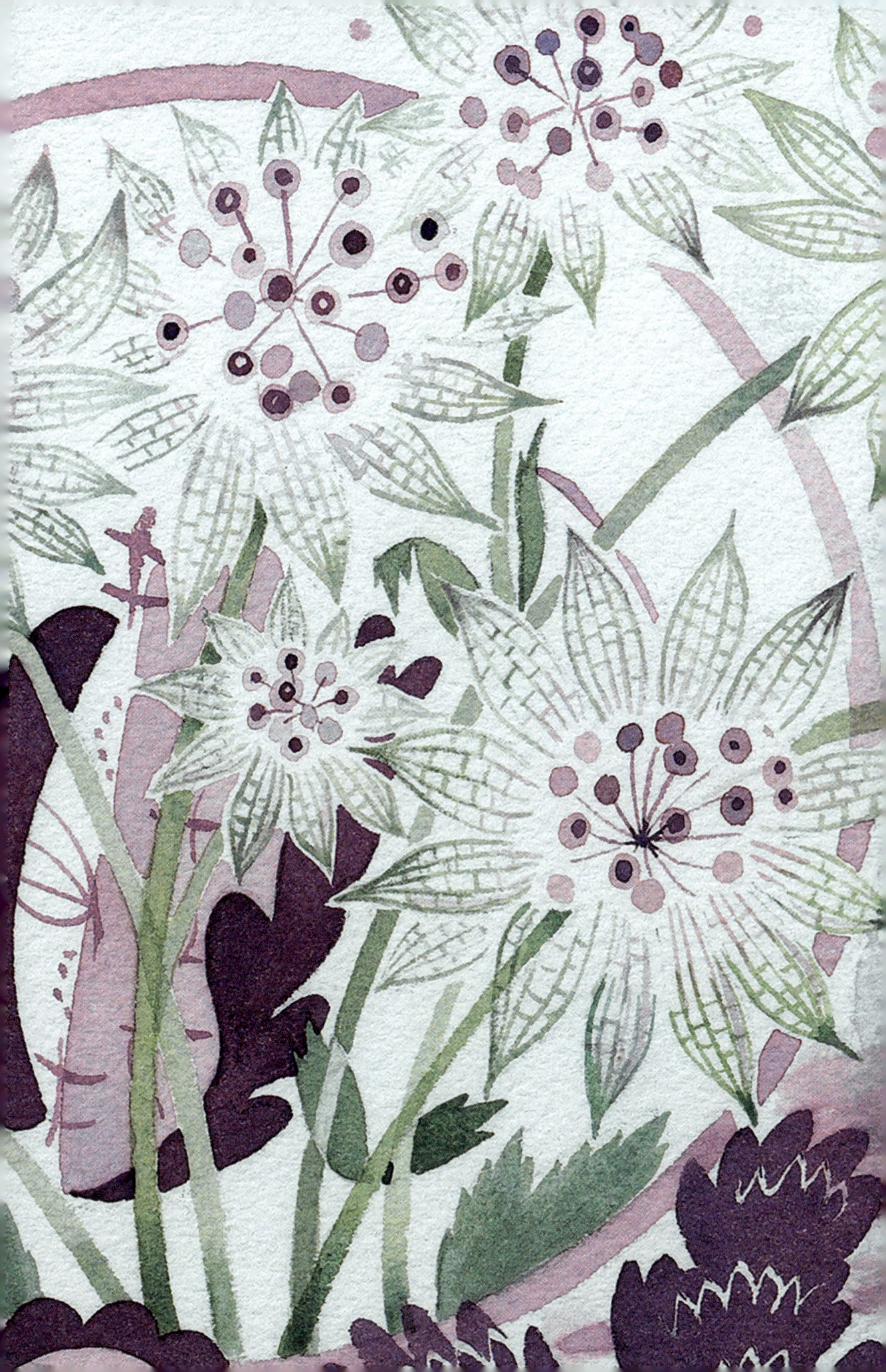

Although most wild astrantias are white or pale pink, horticulturists have seized on the occasional more vividly coloured variants that spring up naturally from time to time. Luckily astrantias produce plenty of seed and are easy to hybridize, so there are some alluring cultivars to choose from. The striking *A. major* 'Shaggy' is characterized by its extra-long bracts. These are greyish-white and shade to green towards their tips, which should be 'toothed' (with several points); if they're not it isn't 'Shaggy'. It's these long, toothed and often slightly twisted bracts that give the flower its shaggy appearance and its name, although it was originally called 'Margery Fish', after the celebrated gardener who first sold it.

Born Margery Townshend in 1892, she started work on Fleet Street and became secretary to six different editors at the *Daily Mail*, marrying the last of them, Walter Fish, in 1933. They bought an old manor house at East Lambrook in Somerset, and over the next thirty years Margery created what became a hugely influential garden, which she opened to the public and wrote about in a series of books. In *Cottage Garden Flowers*, first published in 1961, she described how this 'very interesting and unusual variation of *Astrantia major* is found in cottage gardens in parts of Gloucestershire. The bracts are pale green, about three times as long as in the normal type, and very shaggy. I have never been able to find any name for this truly decorative flower, nor does it seem to be included in any dictionary. I have been given a plant and I hope it will increase and seed itself as generously as the ordinary *Astrantia major* does.' Evidently it did, and in the decades since, 'Shaggy' has become a firm favourite with gardeners.

One of the visitors to East Lambrook in the late 1960s was a novice gardener who, in her forties, had recently moved to a big country house not far away, called Hadspen. Later a sought-after garden designer in her own right, Penelope Hobhouse remembers that Fish gave her a spade 'and told me to go and dig up anything I wanted!' One wonders if some of the plants she excavated were astrantias, because they were always a feature of the gardens at Hadspen, which Hobhouse first opened to the public in 1970. She also opened a nursery with the plant-breeder Eric Smith, who developed numerous well-known cultivars of hostas, bergenias, Japanese anemones (such as 'Hadspen Abundance') and hellebores, including the delightfully named *Helleborus* × *ericsmithii*.

Hobhouse left Hadspen in 1983, but not long afterwards the walled gardens were taken over by a dynamic Canadian couple, Sandra and Nori Pope, who turned them into a place of pilgrimage until their retirement in 2005. Brilliant colour combinations were their big thing, and among the cultivars they developed during their tenure was *A.* 'Hadspen Blood', which has (or certainly should have) rich ruby-red bracts with tightly bunched centres, and often flowers a second time in the autumn. 'Hadspen Blood' is a hybrid between *A. major* and the larger, pinker *A. maxima*, which originates in Turkey and the Caucasus. Although the Popes' once-famous garden is now only a fond memory, their alluring astrantia makes for a marvellous memorial.

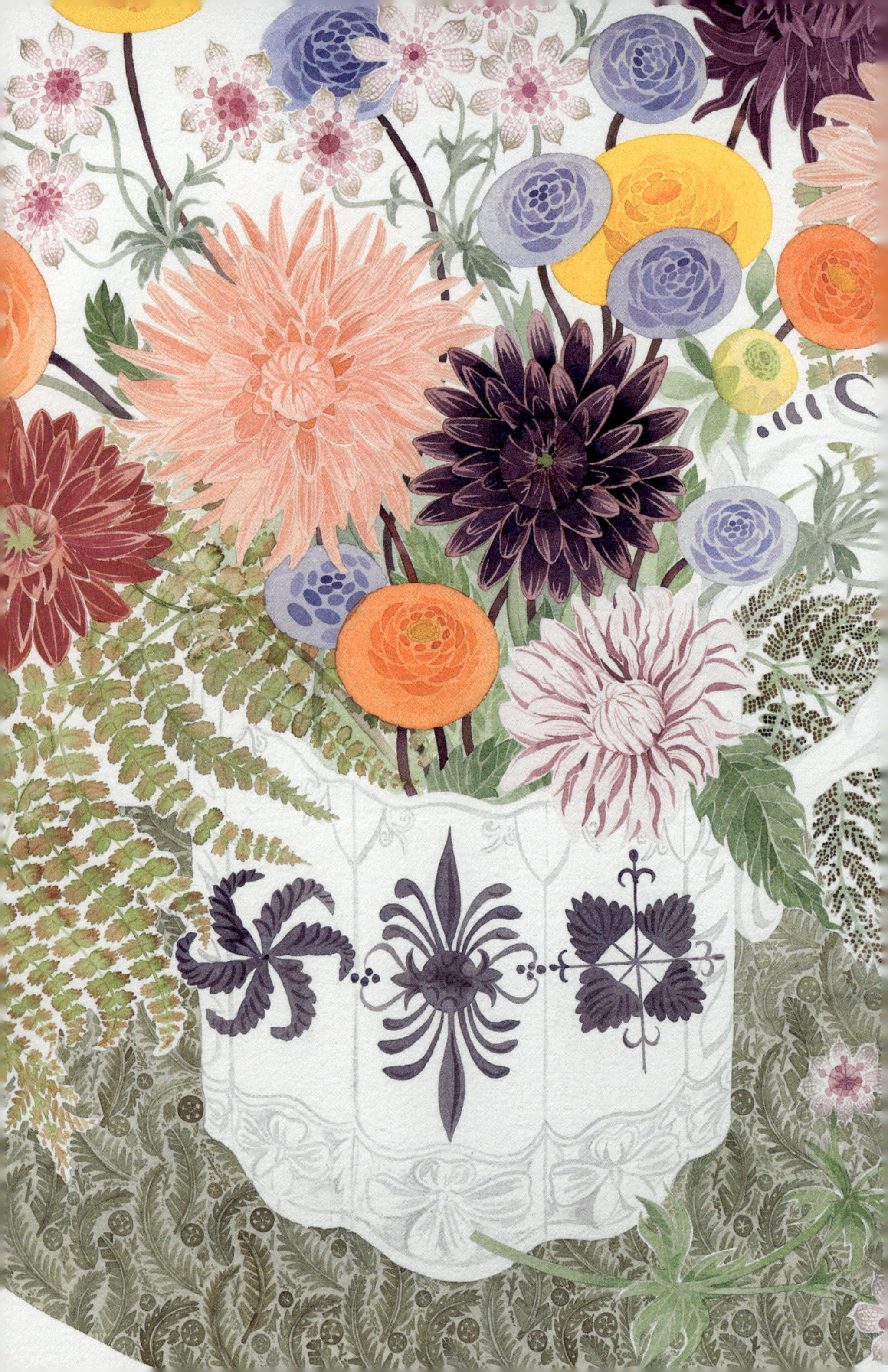

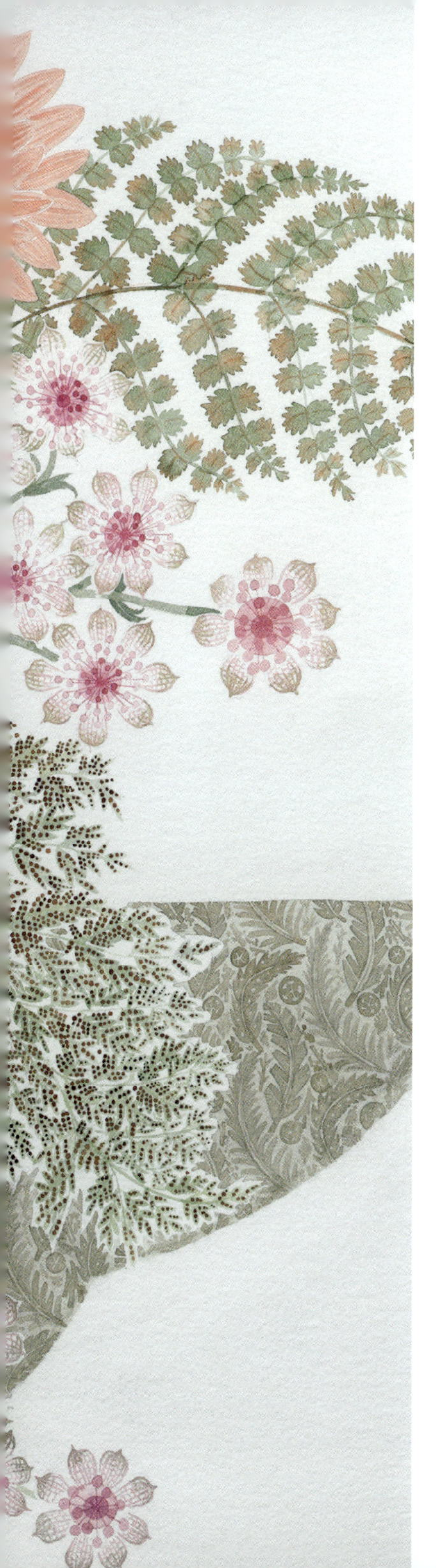

Dahlias, Astrantia and Fern (detail)
Watercolour, 2019

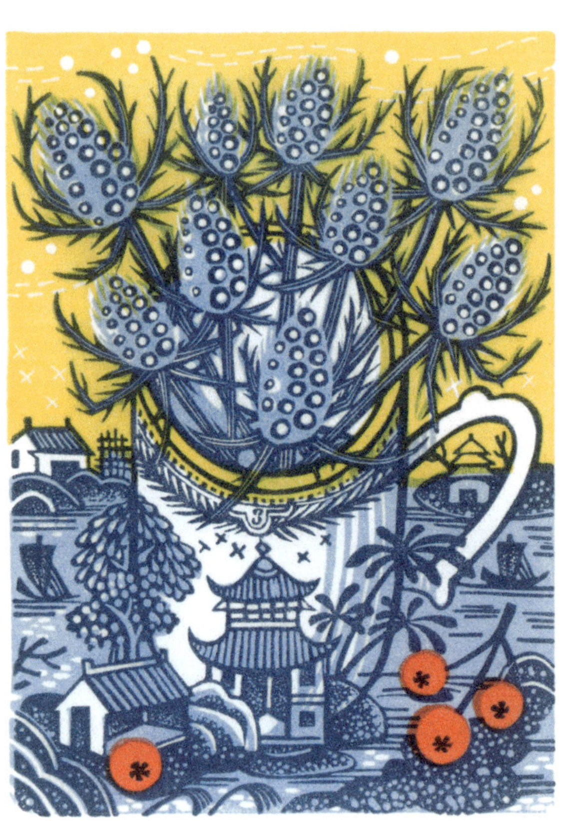

18

ERYNGIUMS

Sea holly is a perfect example of how common names for plants –
while often having the advantage of being easy to remember, poetic
and descriptive – can cause total confusion, in a way that botanical
Latin names, for all their unpronounceability, do not. There are
something like 250 wild species of sea holly to be found scattered
across almost the entire world, from North and South America to
Europe, and North Africa to Kazakhstan, China, Australia and
New Zealand. Yet few of them grow near the sea, and not one of
them is remotely related to holly, although some do admittedly have
prickly leaves. Eryngiums, to use their botanical name, are actually
in the Apiaceae, which means they're in the same genetic family
as celery, cow parsley, carrots and many culinary herbs, including
cumin, coriander, parsley, fennel and dill.

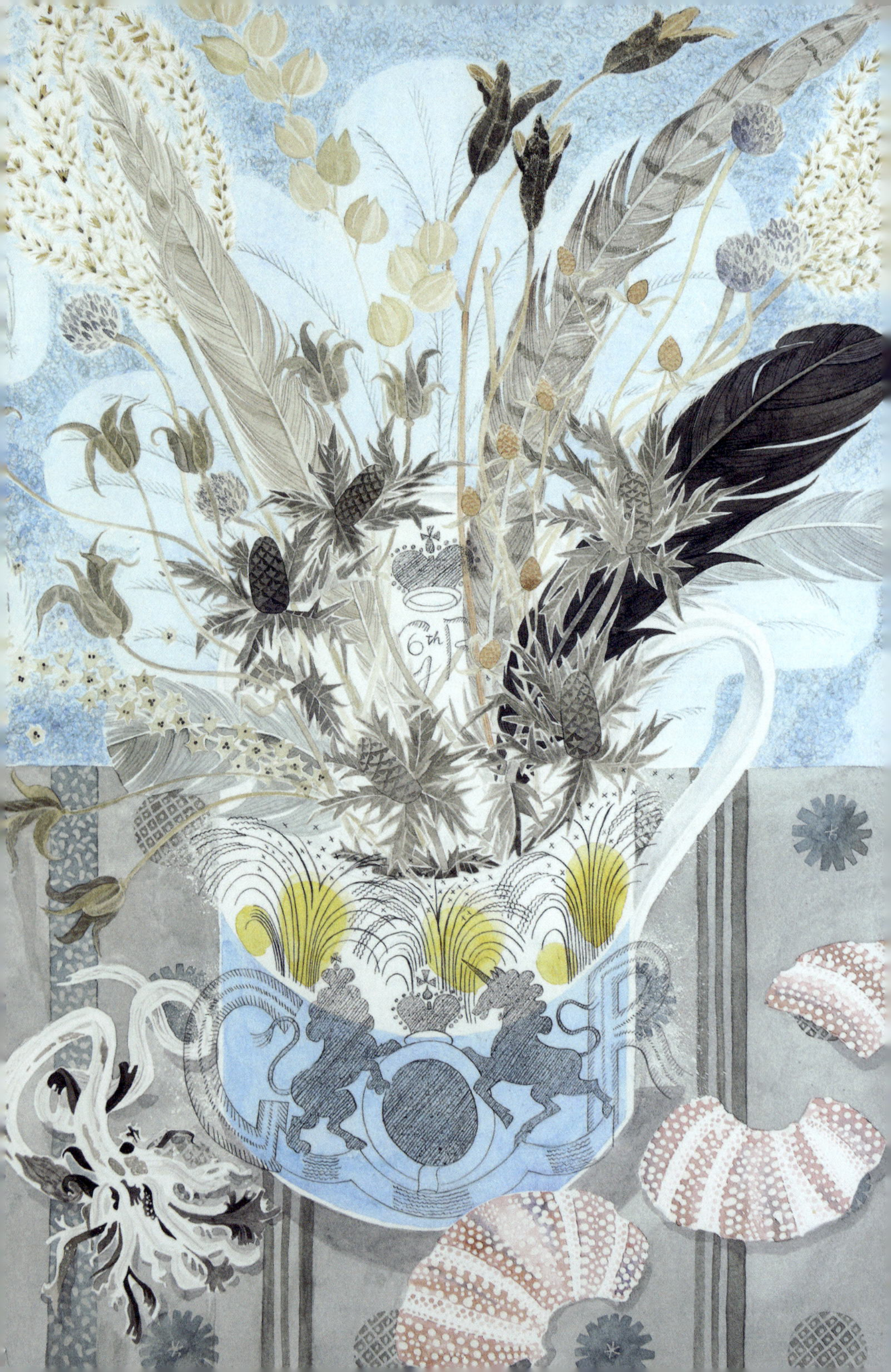

Having said all this, the native British species, *Eryngium maritimum*, really does answer to its common name, being most often found on seashores and having spiny leaves. With their deep roots and leathery leaves, sea hollies are well adapted to high winds and salt-laden air, and in Britain they grow in sand dunes and on shingle beaches all round the coast, although they're more often seen in the west than the east, and becoming less common in eastern Scotland. Their tap roots can be more than 1.5 metres long, which is how they can survive on parched sand dunes, and they're long-lived, too; individual plants can be up to thirty years old. The waxy surface of their leaves helps them to withstand drying winds and strong sunshine – although that also means they need an open, sunny position in the garden.

But, of course, there are hundreds of other species to choose from. With their angular, architectural structure and often blueish foliage, eryngiums are eye-catching plants, so it's not surprising they have had some influential fans. Perhaps the most famous was the remarkable horticulturist Ellen Willmott. The eldest daughter of a prosperous Victorian solicitor, she struck lucky with her god-mother, Helen Tasker, who died in 1888 and left Ellen and her sister multimillion-pound fortunes. The money had been made by Tasker's father, Joseph, a City banker who did very well out of the United Mexican Mining Association.

Ellen Willmott had begun gardening with her mother and sisters on her father's country estate, Warley Place, near Brentwood in Essex. She was ambitious in her plans, even before receiving her huge inheritance. At the age of twenty-one she oversaw the crea-tion of a 1.2-hectare alpine garden, complete with artificial ravine, waterfall, glass bridge and filmy-fern cave. But once she had really come into the cash, her plans – and her spending – started to get out of control. She bought houses with large gardens near Aix-les-Bains in France and Ventimiglia in Italy (the latter also came with a kilometre-long private beach), but it was at Warley Place that she really went to town. It's claimed that she grew something like 10,000 different species and cultivars, and employed over a hundred gardeners, who kept everything immaculate under her gimlet eye.

Willmott planted in a largely naturalistic style, inspired by the writings of William Robinson, with drifts of spring bulbs and huge herbaceous borders featuring hardy perennials, including eryngiums. According to legend, she became so enamoured with one kind in particular – *E. giganteum*, a tall, beautifully silvery species from the Caucasus mountains – that she surreptitiously scattered its seeds in other people's plots, giving it the nickname 'Miss Willmott's Ghost'. It's a great story, but, as with so many oft-repeated tales, it seems to have been made up as late as the 1980s. What we do know is that, sadly, Ellen spent all her money on plants, plant-hunting expeditions and other leisured pursuits, and she began to run seriously into debt. In 1907 her French villa burned down, and before long the house in Italy had to go, too, along with her legions of gardeners.

Willmott spent her last years in solitude at Warley Place, becoming increasingly eccentric and paranoid, booby-trapping her bulbs and carrying a revolver in her handbag, while her once-spectacular gardens grew increasingly unkempt and overgrown. She died there alone in 1934, and the estate was bought by a developer, who

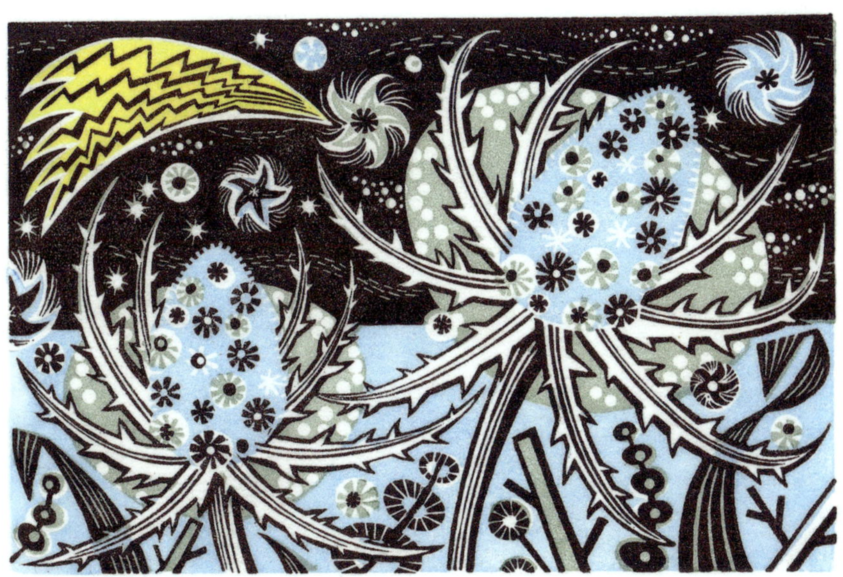

demolished the Queen Anne house in 1939 in favour of an ill-fated housing estate. The houses were never built, the land reverted to nature, and today the remaining 9.7 hectares are run as the Warley Place Nature Reserve by Essex Wildlife Trust.

Eryngiums might then have fallen from favour, but their fortunes revived in the 1970s thanks to another great Essex-based plantswoman, Beth Chatto. At her nursery and garden near Colchester, one of the driest areas in the country, Chatto experimented with hundreds of drought-tolerant plants, including eryngiums, which she started showing on her stand at the Chelsea Flower Show. In 1978 she published her influential book *The Dry Garden*, recommending ten different eryngiums, and their popularity has only grown since then.

First Frost
Wood engraving, 2020

Overleaf: *Delft Blue with Sea Holly and Feathers*
Watercolour, 2022

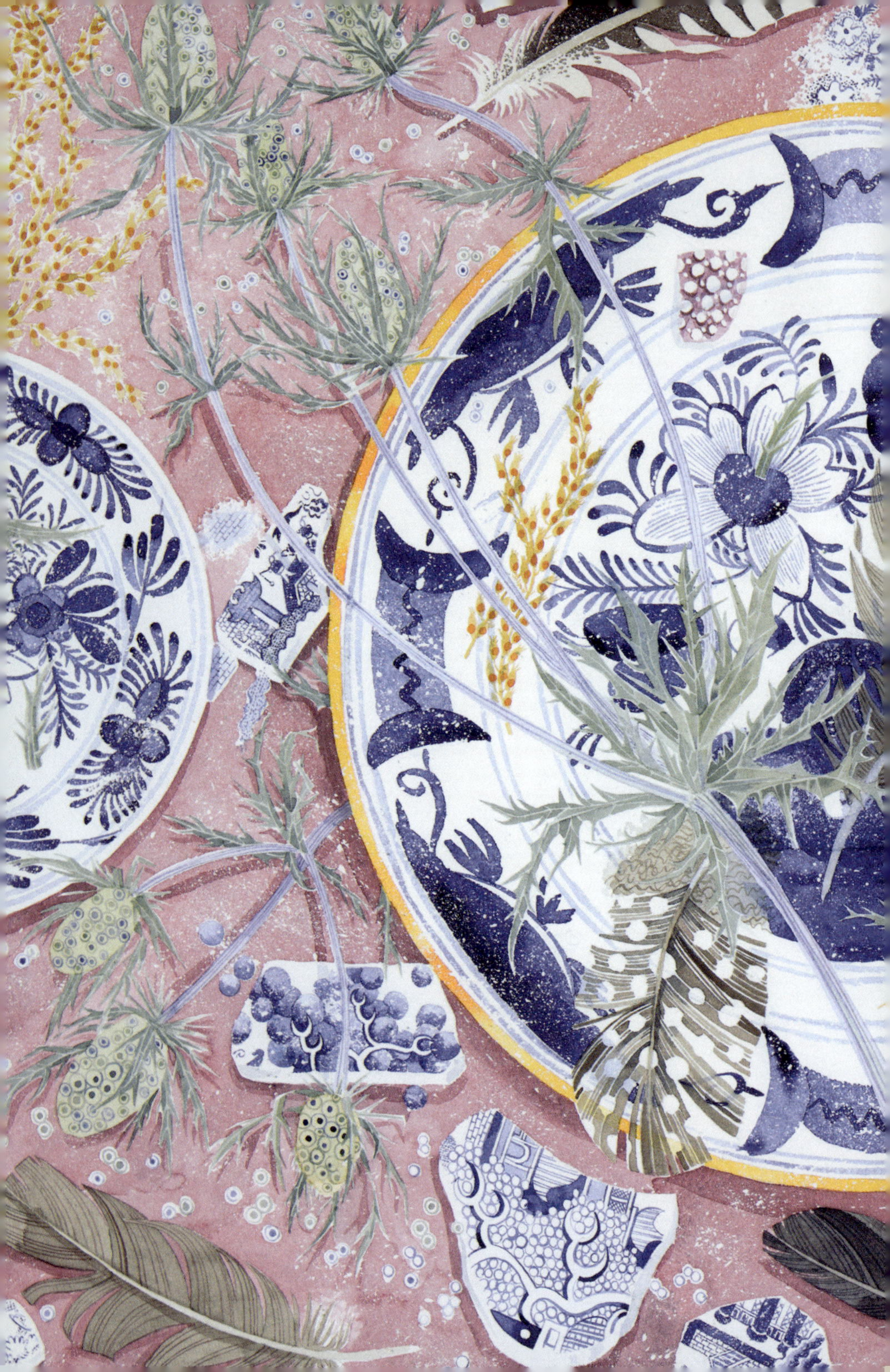

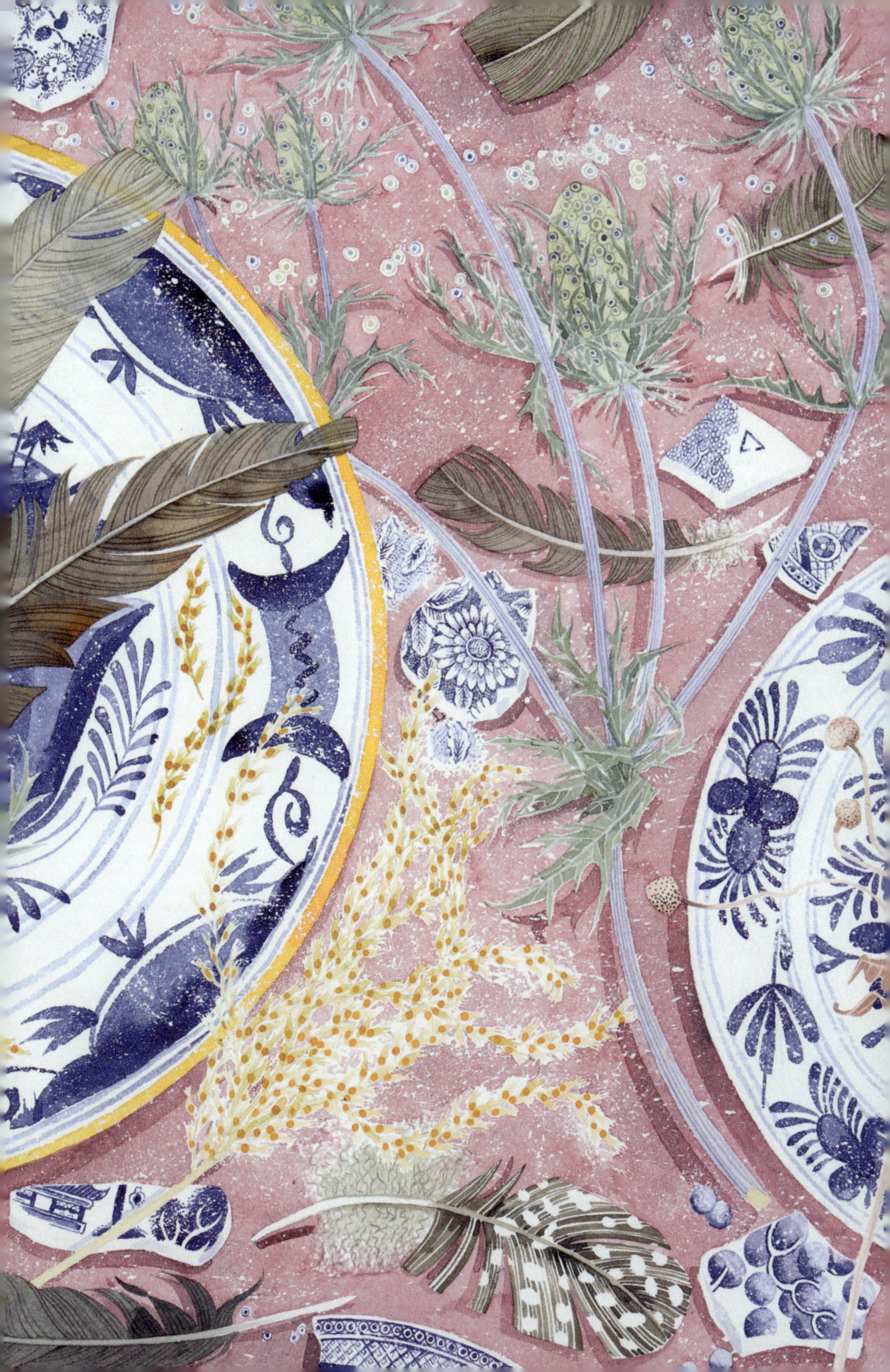

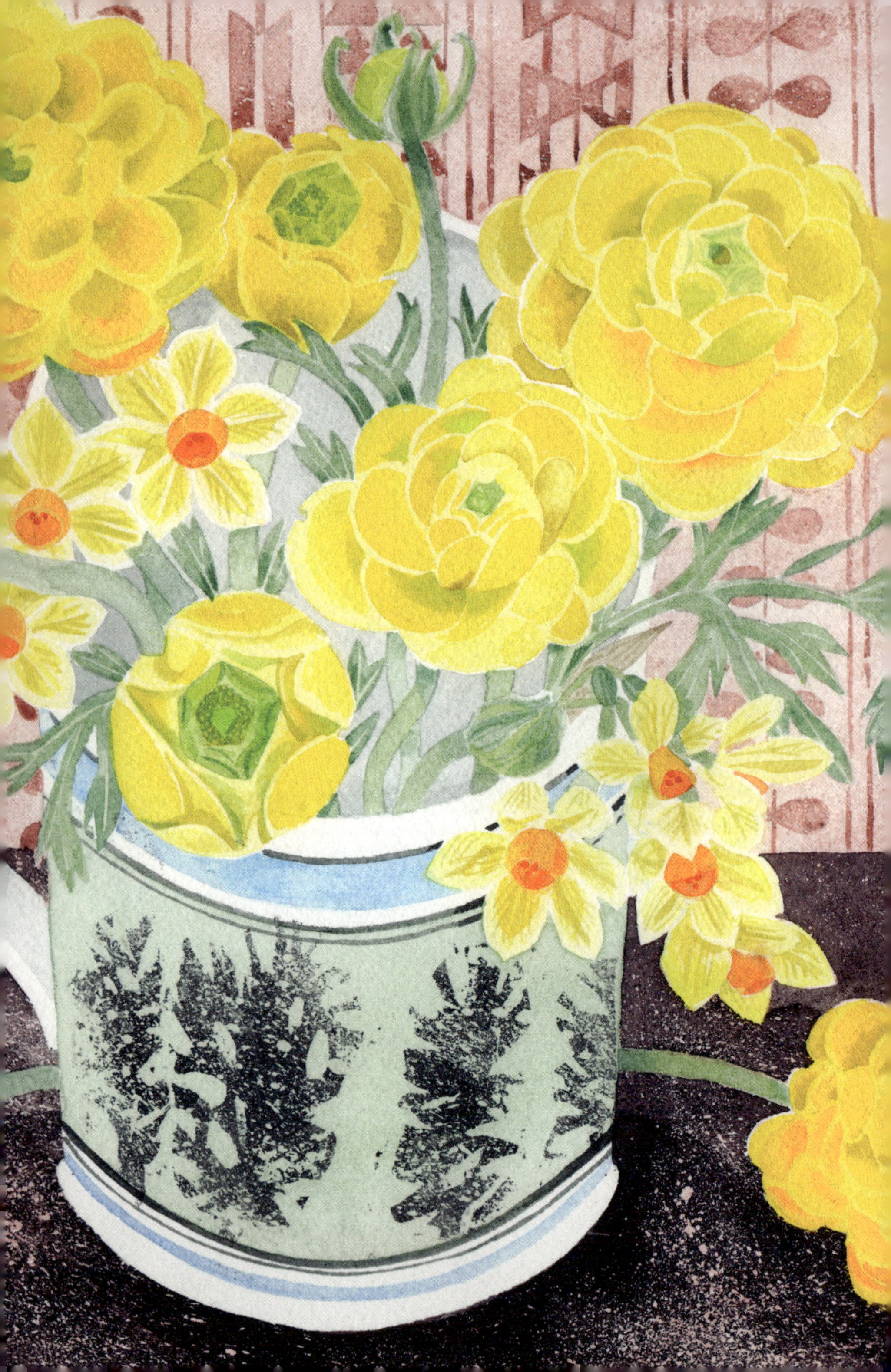

19

RANUNCULUS

Few flowers demonstrate the astonishing ingenuity of plant-breeders – ancient and modern – better than the so-called Persian buttercup, *Ranunculus asiaticus*. It can be seen growing wild around the eastern Mediterranean, from Libya in the west to Turkey in the east, as well as in Iraq and Iran, but in its natural form it looks nothing like the bowl-shaped, many-petalled ranunculi most people will be familiar with as cut flowers. That's because the plain or non-garden *R. asiaticus* looks more like a poppy or the single-flowered *Anemone coronaria*, for which it can easily be mistaken.

Just occasionally, though, a plant would spring up with double flowers as a result of natural mutation, and these were highly sought after by gardeners from early times, partly for their beauty but perhaps even more for their sheer novelty. Many of the areas where they grew were ruled for centuries by the Ottomans, who were highly sophisticated horticulturists, and early double varieties became known in Europe as Turkish or Double Turban ranunculi (or ranunculus or ranunculuses – the name has no set plural form). The Victoria and Albert Museum owns a beautiful watercolour study depicting one of these types, thought to have been painted by the Dutch artist Simon Verelst, who died in 1721.

Sadly these cultivars eventually died out, as did so many others, since they would not have 'come true' from seed and would thus have needed to be regularly regrown from leaf cuttings or root divisions. New semi-double forms were developed in France around the middle of the nineteenth century, and these often had an attractive blotch on their petals, but it wasn't until the early twentieth century that the first really double, 'peony-flowered' cultivars were produced, apparently by Italian growers. This is the kind that dominates the cut-flower market today, and you can see why; they're striking flowers, and can last for ten days or more in water. These early cultivars had one drawback, however: their stems were rather flimsy, which meant they were easily knocked over by rainstorms in a garden setting, and were hard to keep upright in a flower arrangement.

Plant-breeders spent decades working on this problem, eventually producing kinds with shorter, sturdier stems, such as the multicoloured 'Telocote' cultivars raised by Luther Gage in the Californian town of Carlsbad, north of San Diego. Gage named them after the owls that nested on his farm, 'telocote' being a Hispanic-American word for owl. Another notable breeder was

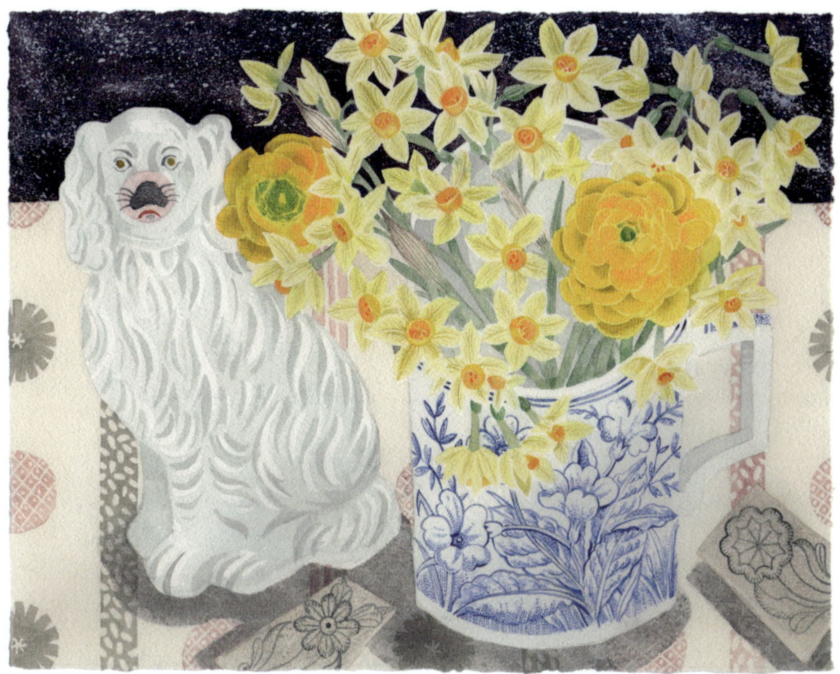

the late Alberto Biancheri, based at Camporosso on the Ligurian coast of Italy, who developed several very successful modern varieties, including 'Elegance', 'Pon Pon' and the self-fulfilling 'Success'. Just to prove the global nature of modern floriculture, another of the most popular multicoloured cultivars, 'Bloomingdale', was bred by the Yokohama-based Sakata Seed Corporation and first released in 1983. Although I have to admit that I still prefer the original wild *R. asiaticus*, no one could deny that the contemporary ranunculus is a triumph of selective propagation, and a testament to the dedication of plant-breeders around the world.

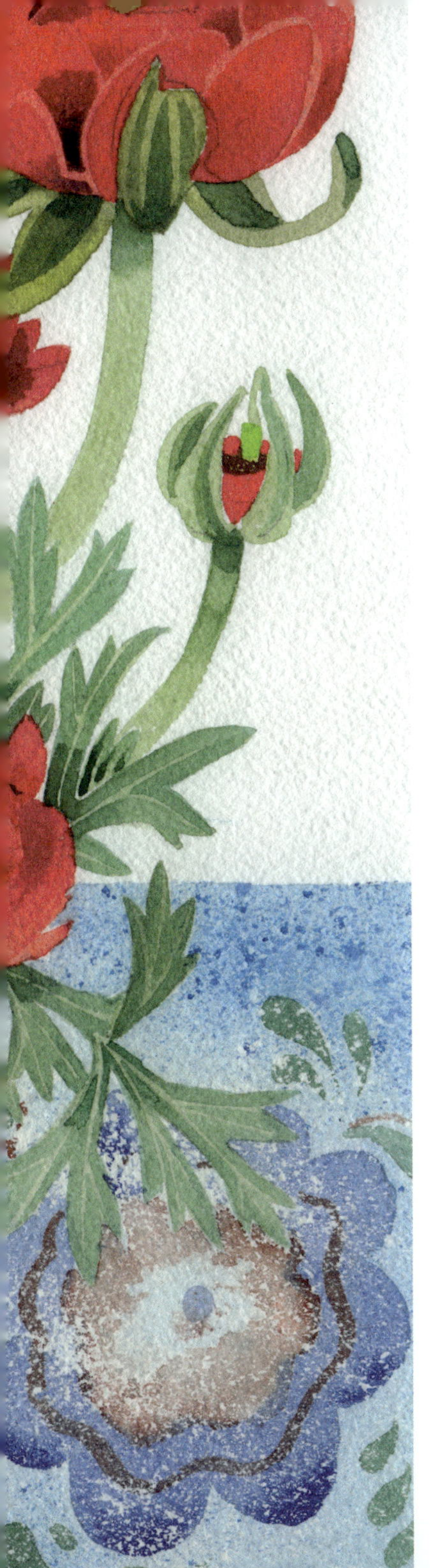

Ranunculus in Gaudy Welsh Cup
Watercolour, 2024

ACKNOWLEDGMENTS

I'd like to thank Nicky Monroe and the Library Enquiries Team at the Royal Horticultural Society, and Dick Fulcher, for their help with the agapanthus chapter, as well as Christine Skelmersdale at Broadleigh Bulbs for her recollections of the agapanthus breeder Lewis Palmer; Nuala Teerink at the Hortus Botanicus Leiden, Arie Dwarswaard at the Koninklijke Algemeene Vereeniging voor Bloembollencultuur in Hillegom, and Robin Stocks for their help with *Anemone coronaria*; Carrie Thomas, former holder of the National Aquilegia Collection; the team at Thames & Hudson; Angie and Simon Lewin for being the best collaborators; and Roy, as ever, for all his help and support.

BIBLIOGRAPHY

Buchan, Ursula: *Garden People: Valerie Finnis & the Golden Age of Gardening* (Thames & Hudson, 2007)
Calendula: An Herb Society of America Guide (Herb Society of America, 2007)
Chatto, Beth: *The Dry Garden* (J. M. Dent, 1978)
Dickens, Mamie: *My Father as I Recall Him* (Roxburghe Press, 1897)

Duthie, Ruth E.: 'English Florists' Societies and Feasts
in the Seventeenth and First Half of the Eighteenth
Centuries' (*Garden History*, Vol. 10, No 1, 1982)
Fish, Margery: *Cottage Garden Flowers* (Collingridge, 1961)
Gerard, John: *Catalogus arborum, fruticum ac plantarum tam
indigenarum, quam exoticarum in horto Johannis Gerardi ciuis &
Chirurgi Londinensis nascentium* (Robert Robinson, 1596)
Gerard, John: *The Herball, or Generall Historie
of Plantes* (John Norton, 1597)
Hyatt, Brenda: *Auriculas: Their Care and
Cultivation* (Cassell Illustrated, 1991)
Johnstone, Katharine: 'Some Cultivated Anemones and
Their Histories' (*Scientific Horticulture*, 1972/73, Vol. 24)
Lawrence, Sandra: *Miss Willmott's Ghosts: A Forgotten
Genius and Her Gardens* (Blink Publishing, 2022)
Nelson, E. Charles: 'Saint Bridgid, Some Daffodils
and a Host of Anemones' (*Northern Ireland Daffodil
Group Newsletter*, Vol. 4, No. 4, April 1994)
Pavord, Anna: *The Tulip* (Bloomsbury, 1999)
Robinson, William: *The Wild Garden* (John Murray, 1870)
Schauenberg, Paul: *The Bulb Book* (Frederick Warne & Co, 1965)

daffodils 70

Dahl, Anders 44

dahlias 2, 7, 10, *12*, 13, 14–15, *16*, *42*, **43–48**, *45*, *49*, 70, 101, *130–31*

Darwin, Charles 15, 34

Dickens, Charles 28–29

Dillenius, Johann Jakob 27

Echinops (globe thistle) 9, 10, *35*, **101–3**, *102*, *103*

Edinburgh 9, *38*

Eryngium (sea holly) *132*, **133–37**, *138–39*, *152*

fennel *112–13*

Fish, Margery 17, 128–29

forget-me-nots (*Myosotis*) 75

Fox, Elizabeth, Baroness Holland 46

France 56, 67, 82, 116, 121, 125, 136, 142

fritillaries *97*

Gage, Luther 142

Gerard, John 41, 54, 116–17, 126

Greece 13, 51, 62, 82, 94, 95, 96

Guatemala 13, 44

Hakluyt, Richard 121

hellebores (Christmas roses) 17, 39, 56, *60*, **61–64**, *64*, 129

Hernández, Francisco 44

Hickman-Windsor, Charlotte 46

Hobhouse, Penelope 129

honesty (*Lunaria annua*) *36–37*, **39–41**, *40*

Hyatt, Brenda 86, 88

Iran 108, 109, 141

Iraq 51, 141

Israel 53, 80

Italy 82, 121, 136, 143

Japan 13, 34

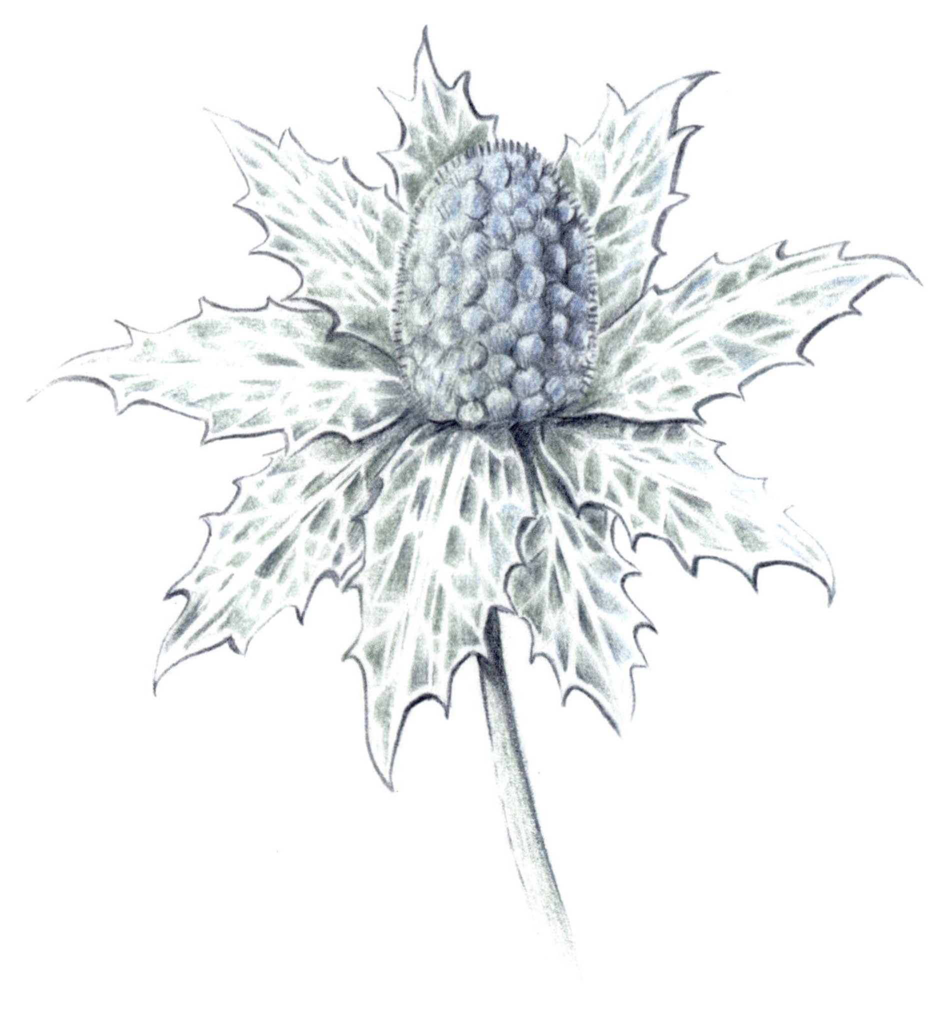

Eryngium giganteum
Pencil crayon, 1997